THE STALINIST COMMAND ECONOMY

The Stalinist Command Economy

The Soviet State Apparatus and Economic Policy 1945–53

Timothy Dunmore, BA (Oxon), MA, PhD

Lecturer in Government
University of Essex

St. Martin's Press New York

St. Martin's Press, Inc., 175 Fifth Avenue, New York, NY 10010
Printed in Hong Kong
First published in the United States of America in 1980

ISBN 0-312-75516-3

Library of Congress Cataloging in Publication Data

Dunmore, Timothy.
The Stalinist command economy.

Bibliography: p.
Includes index.
1. Russia—Economic policy—1946-1950.
2. Russia—Economic policy—1951-1955. I. Title.
HC335.7.D8 1980 338.947 79-26712
ISBN 0-312-75516-3

For my wife Jo

Contents

Tables

CHARTS

MAP

Acknowledgements

My grateful thanks are due to Mary McAuley and Peter Frank of Essex University for their advice and assistance in the past and to my wife for her encouragement and advice during the writing of this book.

1 Introduction

The Command Economy

To realise the huge increases in production called for by the first Soviet Five-Year Plan from 1927–8 onwards, an organisational structure had to be set up to mobilise men and machines and to concentrate human and material resources on major projects. The core of this structure was the people's commissariats (renamed ministries in April 1946). They were to administrate particular branches of industry in accordance with the dictates of the Five-Year Plans. These commissariats were intended to be under the control of those who dictated the plan targets, that is, Stalin and his Politburo. This system by which Stalin sought to control Soviet industry is commonly known as the command economy.

The command economy reached its height during the post-war Stalin era from 1945 to 1953. It was during this period that the commissariats' power over industrial production was at its greatest. Initially the numbers and the power of this state bureaucracy had expanded with increasing industrial output in the 1930s. However, the purges of that decade weakened this managerial elite, but no more than they weakened most other sections of Soviet society. From June 1941 onwards the needs of war dictated even more centralised control over heavy industry for defence needs and it was this centralisation that restored the power of the commissariats. Yet it did not restore the power of many rival organisations, notably of the Communist Party. As a result, by 1945 the officials of the command economy were in at least as strong a position as they had been in the mid-1930s, if not stronger. In addition the war had shown the leadership just how much they needed the experience of the commissariats in the running of a centralised economy; even a capitalist nation like Britain had had to resort to strict central control to direct resources towards the war effort.

It is commonly assumed that this power of the ministries was at the complete disposal of Stalin and the leadership. The noted economic historian Naum Jasny has termed the period 1945–53 as that in which 'Stalin had everything his own way'.[1] The argument of this book, however, is that, in view of the power that the bureaucracy itself possessed, Stalin could not exercise complete control over it nor over the economy. In the command economy it was often the ministries' commands that were obeyed rather than Stalin's, partly because the commands that were published in the form of plans were not merely Stalin's, but were the outcome of quarrels and debates amongst factions within and outside the leadership. An outline of the discussions and quarrels within the Politburo is provided in the next section of this chapter. A more detailed analysis of policy discussion in particular spheres is presented in Chapters 3 and 5.

A further reason for doubting Stalin's authority within the command economy is that the bureaucracy did not and could not carry out all the decisions and directives that the leadership handed down to them. The sheer size and organisational complexity of the command economy led to a degree of confusion that allowed ministerial bureaucrats ample opportunity to distort or even ignore the leadership's plans. (See the third section of this chapter.) A more detailed discussion of the means by which they could do so is provided in Chapter 2. Chapters 4 and 6 show how ministerial officials altered the direction of the leadership's plans in two major spheres of economic policy.

Factional Politics 1945–53

Since Robert Conquest's pioneering study of factional politics within the Soviet Politburo from 1949,[2] several historians have provided further evidence of limits to Stalin's personal authority in the 1940s. Few are now prepared to accept too literally Milovan Djilas's picture of senior Politburo members obsequiously following Stalin about and taking his orders down on a convenient note pad.[3] Stalin was by far the most important figure in the leadership, but he was not the only one that mattered. The period from 1945 to 1948 was marked by continual strife between the Malenkov – Beria faction and that of Molotov and then Zhdanov. After the latter's death and the bloody removal of several of his allies, Khrushchev came to represent the main opposition to Malenkov and Beria. Stalin did not orchestrate all this for his own amusement. He needed the support of some of his Politburo and had to resort to these more devious means of control over them.

Policy debate within the leadership did influence the policy-making process in the Soviet Union under Stalin. For example, foreign policy was a continual subject of conflict within the Politburo. As early as 1944 there were quarrels within the leadership over the question of the form in which reparations should be exacted from the Soviet-occupied zones of Europe.[4]

William McCagg has shown how over the period 1944–8 'insurrectionists' like Zhdanov could push Stalin's foreign policy into a far more anti-Western and pro-revolutionary stance than the 'dictator' wished.[5] Marshal Shulman has argued that the roots of Khrushchev's policy of peaceful coexistence with the West go back deep into the Stalin era to 1949.[6]

In the sphere of cultural policy the change from the 'thaw' of the war and early post-war years to the repressiveness of Zhdanovism in about 1946 was also partly a matter of factional politics.[7] Long after the end of this thaw, but before the post-Stalin liberalisation, 1952 saw further debates over literary policy; not even then were the advocates of liberalisation silent.[8]

In the sphere of agricultural policy the attempt to restore discipline within the collective farms from 1946 was very much associated with the Zhdanov faction which McCagg calls the 'party revivalists'. The 1948–9 rise of the theories of Lysenko and Williams and the 'Stalin Plan for the Transformation of Nature' were pushed by the Malenkov-Beria 'statist' faction[9]. By this time it was Khrushchev who was the main representative of the party interest in agriculture. He has provided us with ample material on his conflicts with Malenkov over the 'agrotowns' proposal of 1951 and Khrushchev's proposals for higher prices for *kolkhoz* produce in Stalin's last years.[10]

In Politburo wrangles in these spheres, as in industrial policy, Stalin is often portrayed as standing aloof, as arbitrating between factions. Yet one recent work argues convincingly that he continually had to boost one faction to counterbalance another, and that the Malenkov–Beria axis in particular was powerful enough after Zhdanov's death to limit Stalin's authority and prevent his attempts to break their power by purges like the Mingrelian Affair.[11]

As we shall see in later chapters there were debates within the leadership over industrial policy in the 1940s; few would dispute this. Yet the evidence we shall present indicates a much stronger hypothesis – that debates also involved those outside the leadership in the party and state bureaucracies. In other words the commands that the command economy machine is supposed to have carried out were partly the result of the influence of parts of that machine itself on the Politburo.

Bureaucrats and Policy Implementation

Just as few students of the post-war years would assume that Stalin's control over his Politburo was absolute, so few would now argue with the proposition that some of the leadership's commands were not carried out by the bureaucracy. Jasny admits that 'What appeared to be a dictatorial power of the highest order proved too weak to enforce proper action even from the ministries and their chief administrations right there in Moscow.'[12] What he is referring to is the fact that the Soviet system of industrial administration did deviate from the Weberian ideal type of bureaucracy in which all orders from above are carried out. Stalin's command economy machine lacked the clear chains of command and 'systematically ordered authority positions' of a Weberian bureaucracy. The responsibilities of the various different hierarchies (state, party, police, trade union, Soviets and the like) overlapped and it was (and is) by no means clear whose orders were to be obeyed. That the orders came from different hierarchies (and often from different sections of the same hierarchy) is due to the fact that the leadership deliberately set all targets too high. It was simply not possible for a bureaucrat or a manager to meet *all* the demands that were made on him. If he managed to meet his target for quantity of output he usually had to underfulfil his 'quality plan' and/or engage in illegal procuring of supplies to do so. High and incompatible targets were set partly always to give an 'incentive' to managers to do better. They were also set to enable the leadership to vary its objectives without revising all of its written plans and directives every time it wanted to do so. By means of leadership campaigns in newspapers and at public meetings and the like it was possible to stress quantity at one time and quality at another, to emphasise one sort of target at one time and one at another.[13] This enabled the leadership to respond quickly to changed circumstances. A bottleneck in, for example, the railway sector could be unblocked by calling on the relevant officials to 'pay greater attention to the transportation of goods'. There was no need to revise the planned use of transport facilities (and thus many other targets) in the annual plan. This 'priority sector' view of the command economy admits that some plans were not fulfilled and some commands not obeyed, but emphasises that the leadership was still firmly in control. It could realise its *basic* objectives if not all of its less important aims.

We argued that the crude Stalin-dictator model of Soviet economic policy-making is now qualified by the existence of conflict within the Politburo. Similarly the crude command economy model of policy implementation must be qualified by the admission of parallel bureaucracies and multiple and conflicting demands. We further advanced the stronger hypothesis that conflict was not confined to the Politburo but spread into the ranks of the bureaucratic machine itself. In the sphere of policy implementation we shall present evidence to suggest that the chains of command were so unclear and silted up that the basic decisions on priorities were often left to the bureaucrats themselves. In particular the leadership could not enforce its most basic priorities of resource allocation due to the power the system effectively allowed to the production branch ministries.

In the next chapter we will try to show *how* this was possible. An analysis of both the legal authority vested in the ministries and rival organisations and the commonly accepted practices of the 'command' economy shows how these 'administrators' could affect the formation and implementation of leadership commands. In subsequent chapters, by means of two case studies, we shall try to assess the *extent* to which they did so: namely, how far plans (and other 'official' decisions) reflected bureaucratic pressures and how far they were fulfilled by the bureaucrats, and whether bureaucratic influence in both respects was limited and sporadic or extensive and biased in particular directions dictated by the bureaucrats themselves.

2 The Organisation of Control in the Command Economy

Organisational Structure

A cursory glance at the Soviet state administration system of the 1940s would reveal a simple line command structure of leadership as set out in Chart 2.1. Such a structure leaves no room for the parallelism that facilitates the emergence both of multiple and conflicting demands on bureaucrats and of divergent interests amongst them. These emerged because even within this apparently simple line structure parallelism existed in the form of a variety of branch and functional administrations (*glavki*) within each ministry and in the variety of ministries themselves. Each trust or enterprise had to deal not with one ministerial master, but with several glavki and with other ministries besides their direct superiors. In addition to this line apparatus there existed a whole range of staff agencies within the state administration framework, from state committees

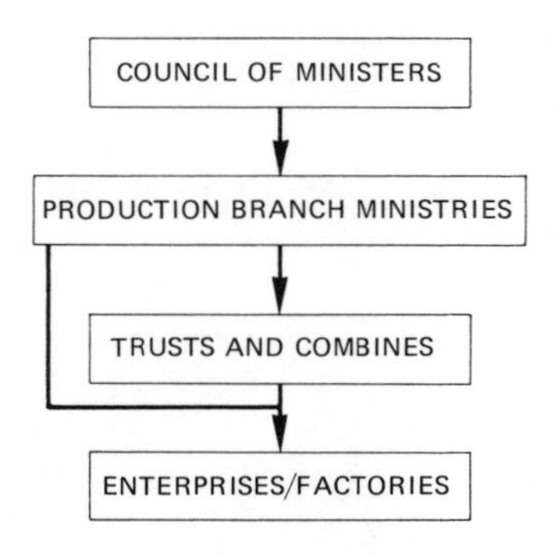

CHART 2.1. Structure of direct control over Soviet industry (simplified)

of the Council of Ministers to advisory bodies at ministerial, trust and factory level.

Generally, within the ministry each official had to deal with a branch-territorial glavk (sometimes through a trust or combine), several functional glavki (dealing with, for example, transport, supplies or marketing), and sometimes staff bodies like the minister's collegium or his technical council. This structure is typified by that of the Ministry of Ferrous Metals in 1947 presented in Chart 2.2. Outside the branch ministry the official had to deal with many other bodies within the Council of Ministers framework. He had to deal with economic functional ministries like those of finance or transport, as well as other branch ministries with which his enterprise(s) had to trade. Thus the Ministry of Ferrous Metals must have had to develop close contacts with the Ministry of Coal and that of Heavy Machine-Building, for both provided for the basic needs of iron and steel works. Further, on some matters the manager and the minister had to deal with non-economic ministries, like that of internal affairs (which controlled the political police). Besides orders from all these quarters the ministerial bureaucrat and the factory manager also had to contend with the staff bodies of the Council of Ministers—the co-

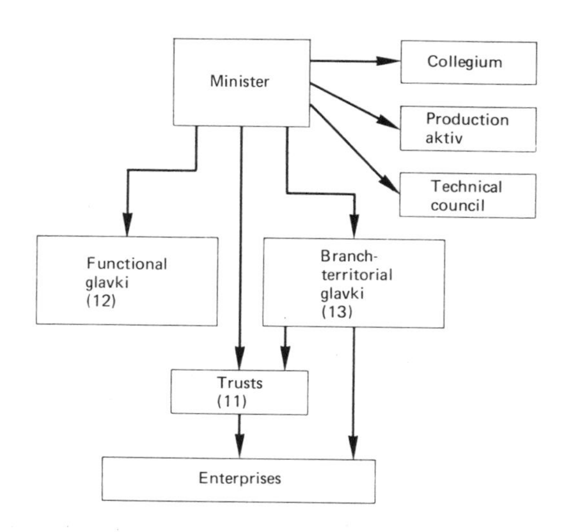

CHART 2.2. Organisational structure of Ministry of Ferrous Metals in 1947*

Source: based on A. Arakelyan, *Industrial Management in the USSR*, pp. 108–10.

* Arrows denote lines of subordination

ordinating State Committees like *Gosplan* (the state planning agency) and (from 1947) *Gossnab* (the state supply committee).

Whilst direct administration of the factory was formally the responsibility of the ministry and the branch-territorial glavk, many other state organisations inside and outside the ministry had responsibility for and made demands on the plant. In addition there existed a mass of controlling agencies outside the Council of Ministers framework. Most important of these were the party, the Soviets and the trade unions. The national and local organs of the Soviets and trade unions had some responsibilities for checking on the activities of ministries in certain spheres like housing provision and industrial safety. The responsibilities of the Communist Party's apparatus were much wider. In effect they were supposed to be the alternative eyes and ears of the leadership within the ministry and the factory. Their responsibility for *kontrol* (checking or supervision) and their organisation along parallel lines to the state apparatus (branch and functional departments) led in effect to their being responsible for all of the ministries' work. The apparently simple structure of the command economy was therefore highly complex and provided ample opportunity for the influence of divergent interests on the formation and implementation of Soviet economic policy. Yet even within the simple structure ministries had enough formal and informal authority to exert such influence.

The System of Direct Control: I The Formal Powers of Ministries

As has been suggested in the previous chapter the simple conception of the command economy relies on the Weberian ideal of bureaucracy. In this orders are handed down from one level of the hierarchy to the next and all the bureaucrat has to do is to carry them out in accordance with an established set of procedures. Modern bureaucratic theory, however, stresses the effect of 'intelligence processes' on the administrator's behaviour.[1] Even if the Soviet bureaucrat had been faced with clear orders and uncomplicated chains of command, he would still have been more than a mere cog in a machine by virtue of the fact that he provided some of the information and advice upon which his superiors based their decisions and commands. In the sphere of economic policy the policy-forming role of 'administrators' in Gosplan, the ministries, and other organisations was enshrined in law.

The general directives upon which annual plans were based were issued in July or August of the preceding year.[2] In 1937 these directives had consisted of only twenty or thirty basic target figures and written notes on policy objectives over the forthcoming year. In the preparation of these directives officials of Gosplan's branch, regional and functional departments were encouraged to air their views.

After these general directives had been approved by the Politburo and the Council of Ministers they were sent to the individual ministries and union-republic Councils of Ministers.[3] Their task was to draw up draft plans for their own sectors or regions in the light of the general directives and send them to the Council of Ministers and Gosplan before 15 November. Gosplan then had a month in which to comment to the leadership on these detailed breakdowns. Only in January (if it was on time) was the plan finally approved by the Council of Ministers. This process of exchanging drafts of plans was essentially one of bargaining, especially between enterprise and ministry or between ministry or region and Gosplan. The aim of each lower participant was usually to secure the easiest output target and the highest input target for his enterprise, sector or region; the aim of the higher official was to set the hardest output target and the lowest input target that was feasible. Managers and ministerial officials frequently 'underestimated' the reserves and other resources at their disposal in order to induce higher authority to give them a lower output target. This and other subterfuges were the result of Gosplan and the Council of Ministers relying for information as to what targets were possible and what resources were needed by each branch or region on those officials who had to fulfil the targets by using the resources. The implementors of economic policy, the ministries and enterprises, influenced the formation of that policy through their provision of the information on which it was based. There were other sources of information available to the leadership, such as local Gosplan and party officials, but they often lacked the resources to check the ministries' figures and sometimes the will to contradict them.

In the implementation as well as in the formation of economic policy ministries in particular had always been allowed some legal freedom. The tasks they set their subordinate industrial plants were governed only in general terms by directions from higher authority. The ministry received overall target figures for its branch from the Council of Ministers and distributed them amongst its enterprises. The limits to this were set by the branch plans and the need to secure final Council of Ministers approval for the distribution. However, the territorial branch glavk set so many targets for each of its enterprises that it was difficult for even the minister, let alone Gosplan and the Council of Ministers, to keep a close eye on all the details. The glavk specified (for each enterprise):

1 The type, quantity, quality and assortment of output to be produced.
2 The amount of labour to be used, wages paid and productivity norms to be attained.
3 The unit cost levels (or cuts in them) to be achieved.
4 The enterprise's allocation of scarce materials for production.
5 Its initial allotment of fixed and working capital.[4]

The leadership clearly could not oversee such details of plan setting, let alone the implementation of those plans.

During the war the GOKO (State Committee on Defence) had kept a close watch on such detail in certain enterprises and branches vital to the war effort, with enterprise directors reporting to it as frequently as once a week. Such a degree of centralisation was not necessary or very functional in peacetime when priorities were more numerous and complex. Yet complaints about overcentralisation in Soviet industrial administrations persisted into the post-war era, and are to be found in many later Soviet commentaries on it.[5] These complaints, however, concentrate on petty and unnecessary supervision over the enterprise and its director by the ministry and its glavk, not by the leadership over the ministry.[6] There was a general trend after the war towards decentralisation of control over the economy, but only to the level of the ministry not the enterprises. For example, a decree of August 1946 removed the requirement that ministries should submit plans for Council of Ministers approval every quarter.[7] Under the new system only annual plans had to be approved (albeit with quarterly divisions). Gosplan and the Council of Ministers were no longer expected to delve into the *minutiae* of ministerial business, provided that annual targets were substantially fulfilled. In addition ministries were allowed to set their enterprises extra monthly output targets provided that these did not increase the total quarterly target by more than 1·0 per cent. They were thus given some leeway to change selectively the plans approved by the Council of Ministers.

Greater ministerial discretion was also allowed in the capital market. Overall capital work plans were handed down to the ministries, but only for the largest construction projects were performance criteria set in the government's plan. A *titul'nyi spisok* of these 'overlimit' projects formed part of the annual plan for the ministry and listed projects of an estimated cost of, for example, over 5 million roubles in machine construction and

10 million roubles in ferrous metals.[8] For these construction sites the ministry was empowered to cut financial provisions to any project by up to 10 per cent and to cut equipment and installation work allocations by up to 6 per cent. By implication, the ministry could redistribute such funds and resources to other projects.[9] In the case of 'underlimit' projects investment funds and resources were distributed by the ministry subject only to overall branch targets set by the government. In addition, of course, ministries and enterprises could finance investment out of their own profits and reserves. However, even after the price reforms of 1948–9 allowed more enterprises to make profits, only about one-fifth of capital investment could have been self-financed.[10] In general, the ministries had a statutory right to distribute and redistribute investment funds to a significant extent, although only within their own production branch.

The labour market for most branches of heavy industry was nominally under the control of the Ministry of Labour Reserves from June 1947.[11] It was they who were to organise the recruitment of labour for industry and train and direct new workers towards the appropriate branches and areas, as specified in plans. Presumably the ministry then distributed the available workforce amongst its enterprises, governed only by plans and decrees relating to its regional distribution. As we shall see in the next section, ministries in fact did much of their own recruiting; and they had the formal powers to offer incentives to attract scarce labour power in a relatively free market. According to the Fourth Five Year Plan 90 per cent of the housing to be built by state organisations was to be built by ministries and their subordinates. It was they, not local Soviets and certainly not the Ministry of Labour Reserves, who could attract workers by offering them living accommodation in war-devasted areas, although they did not always fulfil these plans.[12] Wage and salary levels were in theory centrally determined. In practice, as we shall see, official rates were often ignored and anyway meant very little unless backed up by supplies of consumer goods. The Ministry of Ferrous Metals was probably not alone in having a glavk for the procurement and distribution of consumer goods, as well as one for housing. The main aim of these departments was to attract workers to the ministry's enterprises.

The market for material and technical supplies (the factor market) was also by no means as centrally controlled as the command economy ideal would have us believe. In fact Gosplan and (from 1947) Gossnab drew up plans and material balances (with the approval of the Council of Ministers) only for scarcer or strategically more important types of output, known as 'funded' production. The number of different types of output controlled by central plans doubled between 1940 and 1953.[13] Yet unfunded production still formed as much as 50 per cent of a ministry's total output, although 30 per cent was a more typical figure.[14] The distribution of unfunded production was thus largely a matter for negotiation between ministries or enterprises themselves. In addition central plans only fixed maxima for allocations of funded materials.[15] It was still up to the ministry and its enterprises actually to acquire the materials they needed; and only very general limits were set as to how they in turn were to distribute their final output amongst potential customers. In this perhaps lay the greatest power of the ministries. In an economy where supply shortages were a far more important constraint on decisions than the need to satisfy demands, ministries, and especially their marketing and procurement glavki, had wide powers over the physical distribution of outputs and inputs. The way in which these glavki actually worked will be described in the next section.

The provision of transport facilities was one of the most tightly controlled markets in the Soviet economic system in the 1940s. Special 'political departments' were established on the railways in September 1948 to keep a close watch on this bottleneck on behalf of the leadership. Since August 1946 Council of Ministers' approval had been necessary for monthly plans in this sector (approval of annual plans only was the rule elsewhere). Yet in the face of such nominal centralisation many ministries controlled some of their own transport glavki (although some of their work was directing intra-works transport.) This illustrates the general tendency of production-branch ministries towards 'empire-building'. They sought to provide as many of their own services as they could, from consumer goods to construction work, labour, and transport, to reduce their reliance on outside suppliers. This tendency rendered the ministry much more independent of control through the market. Its formal powers ensured it a degree of autonomy even in theose areas where the market nominally gave way to centrally planned directives.

The System of Direct Control: II The Informal Behaviour of Ministries

The multiplicity of plan targets and the high levels at which they were set (to provide an incentive to raise output) allowed ministries and lower organs a far greater degree of autonomy in practice than was laid down in the formal rules just outlined. So many different and exacting commands were given to the ministry that it could not fulfil them all. Indeed, it seems that it was expected to break some of these rules and ignore some of these commands in order to obey at least some of the others.[16] The only way that a ministerial official or an enterprise manager could do his job at all was to ignore some of the formal limitations on him. This was probably that much simpler if he was not even sent his plans in time to carry them out, or if they were altered during their currency, as often happened.[17]

Many firms went so far as to falsify their reports to higher authority on plan fulfilment. *Pravda* often expressed its editorial concern about such falsification.[18] For all those officials caught in this act, many more got away scot free, and probably still more relied on techniques for hoodwinking superiors that were not quite illegal, such as lowering the quality to raise the quantity of output, adding from stocks to improve output figures, inflating aggregates of different types of output by the use of unrealistic prices, and so on. Many firms and ministries, however, simply admitted that they had not fulfilled all their assigned tasks but hoped to get by fulfilling (or even overfulfilling) a few major targets (often the quantity of output of their main products). This *de facto* freedom in choosing which targets to fulfil was matched by a significant degree of autonomy in the way in which inputs were procured. If an organisation obeyed all the rules and plans governing its inputs, it would probably never fulfil any of its output plans. To achieve the latter it had to indulge in behaviour that was usually frowned upon from above and was often frankly illegal.

In the sphere of capital construction, work often went ahead without any proper estimates of costs being drawn up. Even where agreements between construction organisations and glavki were drawn up they were frequently not fulfilled on time, if at all, and expenditure of funds and resources was often well above the agreed level.[19] The mere existence of a central plan directive for the construction and equipping of a new factory was no guarantee that it would be built. The construction trust concerned could not usually fulfil all its tasks and so had to choose to fulfil some rather than others. The result was the non-fulfilment of plans for many over-limit projects. A decree of May 1950 noting this fact criticised the 'uncontrolled use by ministries and organisations of a significant proportion of the resources directed towards capital work'.[20] The production-branch ministries had to divert resources by whatever means available to get any construction work at all done for their enterprises.

The labour market was probably even more independent of central control than the capital market. A Council of Ministers decree of November 1951 accused the Ministry of Labour Reserves (and also the executive committees of local Soviets) of lacking control over the labour recruitment process and of simply not recruiting enough workers. The result was that ministries and enterprises were illegally sending their own officials to collective farms to recruit new workers.[21] In so far as the servicing agencies were not fulfilling their obligations to the production-branch ministries, the latter were forced to ignore central directives and take matters into their own hands.

The gap between legal provisions for a tightly controlled market and the near anarchy that actually prevailed is well illustrated by the case of the 1940 labour laws. These had effectively prohibited a 'self-willed' change of job by any worker and introduced severe penalties for lateness, idleness and so on.[22] In practice, however, very high rates of turnover were a great problem in post-war Russia.[23] Many workers simply left their jobs to seek better conditions and pay elsewhere. The shortage of labour caused by war losses and the return of women and very young and old people to the land after the war resulted in many enterprises hiring whatever labour they could find, irrespective of whether the worker had the correct documentation (including the infamous 'work-book').[24] Centralised direction of labour seems to have been confined to relatively few campaigns aimed at particular trouble spots such as the Donbass coal mines in 1949–50.[25] Such campaigns could, by their nature, not be used in the general case. As indicated above, branch ministries had powers over the provision of housing and consumer goods that enabled them to attract workers. They did not even seem particularly bound by centrally determined wage rates. Very different rates of pay were offered for the same work by different ministries and departments during the 1940s.[26]

The freedom of ministries and enterprises from central control over the procurement and marketing of material inputs and outputs was (and is) epitomised by the ubiquitous *tolkach* ('expediter'). His function was to secure supplies by any means possible. In so far as supplies were provided through the normal channels (in fulfilment of contracts between glavki or enterprises in accordance with central plans) tolkachi were unnecessary. The supply situation in practice was such that one author in 1946 blamed over-high administration costs partly on the employment of tolkachi! He further revealed that an enquiry conducted by Prombank (the industrial bank providing long-term credit) had shown that 5 per cent of the cost of materials went to paying 'ordering officials', well above the 'norm' of 3.5 per cent (which was supposed to cover other expenditures and losses as well!).[27] No wonder many branch ministries sought to supply as many of their own needs as possible.

In general one can only agree with David Granick that 'in practice a commissariat [ministry from 1946] has usually been able to switch about as it wishes the distribution of its supply allotment among its firms.'[28] One could go further and argue that a ministry's supply allotment was often very different from what had been specified in leadership-approved plans. This was the product partly of the formal powers of the ministries and glavki (and particularly those directly controlling the productive enterprises) and partly of the way their officials were forced to ignore some central directives in order to survive. It was also due to the attitudes of those officials. They had the opportunity to selectively ignore directives and they also had an interest in doing so. The aims and interests of ministerial officials were well summed up in A. G. Frank's survey of the aims of industrial managers; that is:

1 their own survival;
2 the expansion of their own 'empires';
3 the extension of their own autonomy.[29]

Survival and expansion dictated the need to attain good overall production figures, but to do so they had to achieve a degree of autonomy from leadership control.

The System of Indirect Control: The Non-ministerial Checking Organs

Yet there existed at this time a series of bureaucracies in the USSR whose organisation paralleled the ministries' and whose task was precisely to check on the implementation of leadership directives by the ministries. That they generally failed to prevent selective non-fulfilment of directives was due both to organisational weaknesses and to their officials often taking the side of the checked upon (the ministry or glavk) rather than that of the checker (the leadership and its checking arms, the central party, Soviet, planning, police, trade unions and other apparati).

This general tendency towards co-operation rather than conflict envisaged by the command model was in large part due to the weakness of those checking organs themselves, notably the party apparatus, the functional ministries, the Council of Ministers' state committees, the Soviets and the trade unions.

The Communist Party Apparatus

Of all these checking organs only the party's full-time apparatus had formal responsibility for all the work of each branch ministry. Other organs were limited (in theory at least) to one aspect of the ministry's work, such as its finances, its planning or the legality of its operations. Until 1948 the party apparatus was itself organised along these purely functional lines. At its head the Central Committee Secretariat had separate departments for, for example, cadres work, 'organisation-instruction' work and so on. Only after the death of the ideologically centred Zhdanov was this structure changed to a more production-branch oriented one. Under the guiding hand of the more pragmatic Malenkov the secretariat established departments for heavy industry, light industry, transport and planning, trade and finance. The larger regional party organisations had a similar branch departmental structure.

The party really lacked the resources, the power and the motivation to challenge the state apparatus and provide an effective check on the execution of leadership commands. In terms of resources the party had a total of less than 200,000 full-time officials.[30] As half of its post-1948 departments at All-Union level were not dealing with industrial matters, one could estimate the number of full-time party officials engaged in checking the branch ministries' work at not much over 100,000. The personnel employed by those ministries must have run into millions; there were 70,000 'ordering' officials in heavy industry alone at this time. The result was that each party official had to check on the work of (on average) ten or more officials and managers. These latter were generally more specialised in their training and their work experience than the party's *apparatchiki*.[31] It is not surprising that a relatively unprofessional party official did not provide a very effective check on ten or so highly trained ministerial officials with long experience of their own specialisms.

The party had a mass membership of about 6 million at this time, but the burden of checking work fell on its full-time officials. This was due to the general lack of contact and communication between the apparatchiki and the membership. As we shall see, party meetings and elections were often not held at all and, where they were, frequently had very little to do with the realities of economic administration.

Yet it was the party apparatus that in theory had the power to bring disobedient state officials to heel. Of the other checking agencies only the police could do more than complain to higher authority about ministerial officials and managers who were not acting as the leadership had specified. The party apparatus had control over personnel selection and deployment (*nomenklatura*) and could bring those state officials who were party members to party discipline. Many important officials in ministries and managers were members of the party but the link between their primary organisations and the apparatchiki at regional and national levels was very weak. Control over personnel was a power that party officials made little use of even after cadres functions were transferred to branch departments in 1948. The party made little or no attempt to remove disobedient officials from the state machinery in the 1940s.

As McCagg has noted, very few leading figures in the state administration were removed from their posts between 1941 and 1953. The group of top state industrial officials around Malenkov was the one faction that Stalin never managed effectively to control.[32] By 1953 the average tenure of office amongst directors of industrial enterprises was as high as ten years, compared to only three years in the 1930s.[33] *Pravda* had called for an improvement in cadres work within ministries and economic organisations in December 1946 but there is no evidence that widespread personnel changes followed.[34]

The decline of the Russian Communist Party's power relative to the ministries had begun in the 1930s and had been accelerated by the purges and the war. The criticisms of party work during the war period (1941–5) concentrate on indiscriminate recruitment policies and poor training of recruits, failure to hold meetings and elections, combining with state organs and engaging in irrelevant theoretical work rather than checking on state officials. The leadership and especially Zhdanov, Kalinin and Molotov were well aware of the dangers of such trends for the post-war Soviet Union. From January 1944 they instituted a 'party revival'[35] aimed at redisciplining its members, improving its work, and replacing the nationalistic propaganda of wartime with a more 'Marxist-Leninist' approach.

However, this revival seems to have had its greatest impact in the spheres of culture and ideology with the infamous *Zhdanovshchina*.[36] At the level of industrial administration its impact was limited. Most of the abuses noted above continued, especially after the death of Zhdanov and the purging of many of his fellow party revivalists in 1948–9 (the 'Leningrad Affair').

During the war the sheer quantity of new admissions to the party had made it impossible to keep all of its members under tight central control. Total party membership rose by 50 per cent during the war in spite of many deaths. During the war 5.1 million Soviet citizens were admitted to the party; its total membership in January 1941 had been less than 4 million.[37] Many were admitted solely for having shown bravery in battle and some new entrants had even previously been expelled from the party.[38] The result was a mass of new recruits with little or no knowledge of party work or party discipline. The Central Committee resolved in July 1946 and March 1947 to restrict admissions to the party. The result was that only 580,000 were admitted to candidate membership of the party over the period January 1947 to October 1952.[39] Even so, careful scrutiny of party membership figures shows that relatively few people were actually expelled.[40] No doubt the mass of wartime recruits were receiving a thorough political training after 1945, but the main problem for the party was the fact that many of its full-time officials had very little time for the mass membership. The failure to convene regular meetings of mass organisations and to hold the 'elections' required by the party rules has been noted even at the highest levels. It is well known that the All-Union Party Congress did not meet between 1939 and 1952 (in spite of strong rumours that the party revivalists hoped to hold one in 1946) and no Central Committee *plenum* was convened between 1947 and 1952. Many local officials followed suit and rarely held committee and general meetings at which they could make contact with the mass of their members and state officials.[41]

Perhaps most significant for this study is the fact that the tendency towards joint meetings and decisions of party and state bodies established during the war continued thereafter. During the war supreme power within both party *and* state had been vested in a State Defence Committee (GOKO) of five and later eight members. At lower levels state and party officials were seemingly encouraged to work together and it was common for the leading party official in a republic or province also to head its Soviet or Council of Ministers. Even after the war Khrushchev, for example, was at the same time both Party Secretary and Council of Ministers Chairman in the Ukraine; G. M. Popov was in 1950 a Central Committee Secretary, a secretary of the Moscow city and provincial party organisations and Minister of Urban Construction. Presumably the latter was supposed to 'check' on himself! That the practice of issuing joint party-state resolutions continued well into the 1950s is shown by the criticisms of this practice at several republic party congresses in 1952.[42] Far from spying on and supervising the state apparatus many party officials worked hand in glove with their state counterparts.

The weakness of the party is reflected in the criticisms of it made at the time and also during the deStalinisation of the Khrushchev era. At the regional level party officials found it difficult to compete with ministerial authority. Apparatchiki were attacked for poor checking work at both the Ukrainian and Azerbaidzhani Party Congress in 1952.[43] In the same year Malenkov accused them of being unable to resist 'all sorts of local, narrowly departmental and other anti-state pulls and pressures and overlooking outright distortions of the Party's policy in economic fields and violations of the interests of the state'.[44] Western studies of the command economy confirm that at both factory and regional level party secretaries played second fiddle to state officials and managers.[45] Far from checking on the latter, party apparatchiki often helped them. One regional secretary accused his heavy industry department of being little more than a 'despatcher's office' for local firms.[46] At least this was better than other party organisations whose general apathy was relieved only by irrelevant 'theoretical work'.[47] Throughout the war and the succeeding decade local party officials tended to co-operate with state agencies in their evasion of central directives.

The central party apparatus did, however, seek to maintain control over a few selected key enterprises by sending its own 'organisers' to them rather than relying on local party officials.[48] Such a grip over the more important sectors of the economy is also reflected in the establishment of political departments on the railways in 1948. These were party organs whose function was to check on the operation of this 'bottleneck' sector of the Soviet economy. Such practices fit in well with the 'priority sector' model of the command economy. The party may have concentrated its energies on looking after only a few priority sectors and enterprises for the leadership. Only detailed studies of priorities can tell us whether party control could cover enough enterprises and branches to enable the leadership to get its 'basic' commands carried out. In general party control through the regional infrastructure was not good enough to secure the implementation of its policies against the wishes of state officials.

Checking Organs Within the Council of Ministers

Within the Council of Ministers the responsibility for checking on most aspects of plan fulfilment by ministries and enterprises belonged to Gosplan. Even after the establishment of separate state committees for supply and technology (Gossnab and *Gostekhnika*) and of a Central Statistical Administration in December 1947, this continued to be the case. Most complaints to government agencies about non-fulfilment of planned tasks eventually found their way to Gosplan.[49] The state planning agency's central apparatus was divided into functional departments for such matters as perspective planning, current ('economic') planning, prices and costs, finance and production. Each major branch of the economy was the responsibility of one section ('special administration') of the production department.

Gosplan's main problems in performing its checking role lay in its lack of executive authority and in its inadequate regional infrastructure. Its officials had no day-to-day control over production matters or personnel and could only report misdemeanours to higher authority. They could not take remedial action themselves. Furthermore, in wartime they had had to spend all their time setting plans rather than checking on their fulfilment.[50]

A decree of August 1946 did end the practice of setting monthly plans for the more important sectors and proportions in the economy and Gosplan's load was thereby eased. However, even after this date, it simply did not have the means to check out complaints of non-fulfilment at ground level in the enterprise or the glavk. It lacked sufficient numbers of regional officials (*upolnomochennye*, or representatives) to do its checking work. Furthermore these representatives were subordinate to other local state organs as well as to Gosplan. Not surprisingly they were often criticised for exhibiting formalistic attitudes and a lack of knowledge of local conditions.[51]

The Soviet secret police also had extensive economic responsibilities. They were, at least in name, under the control of the Council of Ministers and had their own economic empire, as well as responsibility for bringing to light contraventions of the law. Its former powers were reduced from 1945 with the transfer of many of the munitions plants under the control of the NKVD to peacetime production under branch ministries. In addition the NKVD, the internal affairs commissariat, had in 1943 lost some of its functions to the re-established NKGB, the state security organisation. This latter body (soon to become the MGB) was in charge of rooting out, amongst other things, such 'economic crimes' as falsifying reports and engaging in black market trading. The organisational basis of the power of the political police and so of Beria was thus divided. In addition Stalin and Zhdanov had sought to limit the power of the police by appointing a staunch party man and arch-Zhdanovite, A. A. Kuznetsov, to be Politburo 'overlord' for police affairs. However, Beria restored much of his own and the police's power in 1948–9 when he purged Kuznetsov in the Leningrad Affair. However, the power of the police in the 1940s did not go unchallenged. In spite of the existence of police 'special sections' in most large enterprises and institutions, remarkably few managers and ministerial officials were purged. Even the feared MGB and its local officials were not averse to co-operating in the malpractices of state officials. At the highest level the period from 1945 was marked by an alliance between the main champions of the police and of the state economic apparatus – Beria and Malenkov.

The other important organs of indirect control over industry were the finance ministry and (subordinate to it) the major state banks, notably Gosbank and Prombank. If direct control of the ministries and firms was not possible perhaps they could be controlled via the market using financial levers. An economist writing in 1946 specified the following necessary conditions for effective 'checking [*kontrol*] by the rouble':

1 Goods-money relations between state enterprises and organisations.
2 Observation of socialist credit principles.
3 Observation of state-fixed prices.
4 Realisation in money form of all losses caused to other enterprises or organisations by failure to meet contractual and planned obligations.
5 Very strict accounting and correct conditions for calculating expenditures.[52]

Yet, in spite of a continuing campaign for financial discipline and accountability (*khozraschet*) throughout this period, few if any of these conditions were realised by the time of Stalin's death. Financial controls over Soviet industry remained the dream of a few officials and academics rather than a reality of the command economy.

The first problem that the Ministry of Finance faced was that many transactions did not pass through the books at all. Organisations engaging in quasi-legal exchanges at goods often preferred to rely on barter rather than provide written evidence of their deeds in bank accounts and contracts.[53] Neither were credit controls very effective at this time.

Short-run credit (for covering the cost of goods in transit, of seasonal variations and the like) was provided by Gosbank. This it seemed to distribute more or less on the say-so of the branch glavk in charge of the firm. Firms often deliberately went into debit in Gosbank to prevent funds being used to settle their other debts or to pay fines imposed on them.[54] The Finance Minister, Zverev, revealed that in 1947 Soviet industrial enterprises were in debt to Gosbank to the tune of 6 billion roubles and to other enterprises to the extent of 14.7 billion roubles.[55] Gosbank had neither the time nor the resources to check on the indebtedness of firms and so was in no position to provide a reliable source of information for the leadership against the ministry, let alone provide a means of control over it.

Most long-run credits for capital construction were provided by Prombank, the industrial bank. This organisation, like Gosbank, seemed to operate more like a charitable foundation for firms and glavki than a control over them. It was complained that six months after the end of the war Prombank was issuing more than 40 per cent of its credits to construction projects for which no detailed estimates or plans had been produced![56] Later in 1946 it was announced that unless such documentation was produced by 1 April 1947 all credits for the project would be stopped from that date. In June 1947 the head of Gosplan's capital construction department complained that the documentation still had not been produced for two-thirds of these projects and yet they were still apparently being funded.[57] Here is an excellent example of the lack of discipline and financial control in the command economy. The direct orders of a government decree were not being carried out, for to do so would have resulted in either more than a quarter of the USSR's centrally funded capital construction projects being scrapped or in officials spending so much time in the necessary paper work that production figures would have suffered.

As to the other preconditions for financial control listed above, prices were often ignored, fines went unpaid and accounts were far from accurate statements of reality. Where transactions between firms were planned and legal they did not always take place at the centrally fixed price levels.[58] One of the most common forms of falsification was to supply goods of second- or third-grade quality at first-grade prices. The recipient of the goods could of course refuse to accept them or complain to the Ministry of Control or to the state arbitration agency, *Gosarbitrazh*. In either case it was often more productive for the recipient to be grateful that he had received any goods at all. Gosarbitrazh did impose fines on enterprises that failed to deliver goods but all too often these fines were simply not paid.[59] In addition, of course, some firms and glavki never got round to drawing up the contracts to fulfil their planned obligations. Under such conditions strict accounting and exact measurement of expenditure were simply not possible. Even at the All-Union level there is every reason to suspect that any amount of statistical manipulation and downright falsification went into preparing reports both for public consumption *and* to please superiors. For example, how did the Central Statistical Administration manage to claim a 73 per cent increase in gross industrial output over the period 1940–50 when only one major sector (machine-building) managed to exceed such a growth rate and most fell well short of it? Officials at all levels were adept at writing reports that inflated output figures and under-estimated costs. The prevalence of such falsification is shown by the frequency of attacks on it in the editorial columns of *Pravda*.[60] The leadership had no organisation within the Council of Ministers structure that it could rely on to provide a clear and accurate picture of violations of its orders. Without that, how could it prevent such disobedience without resorting to the arbitrary terror of 1936–8? Even that terror had proved economically counterproductive, with Soviet industrial growth rates declining in 1938.

Other Checking Organs

The other major checking organs, the Soviets and the trade unions, were limited in their spheres of competence and very much under the control of party and state organisations at the local level. Local Soviets had authority only over locally controlled industry, some housing and social services; they had no responsibility for All-Union industry. Local industry did account for 27.5 per cent of Soviet industrial output in 1950, but more than half of this was produced by 'local' enterprises under the RSFSR Council of Ministers and Supreme Soviet. These were controlled from Moscow and the RSFSR Supreme Soviet had no more control over or information about individual enterprises than did the All-Union Supreme Soviet. Both lacked officials in the regions. Of course the 1000 or 2000 deputies to these two Soviets who were not senior officials in other organisations could report problems and plan violations in their own constituencies. They were often encouraged to use their own organ, *Izvestiya*, to air such complaints. In similar fashion trade union officials whose concern was to be directed towards working conditions and workers' welfare could make complaints in 'their' newpaper, *Trud*. In both cases only a small selection of complaints ever saw the light of day and ministries frequently ignored them. Soviets and trade unions lacked the 'teeth' and the resources to control the ministries and factories.

Parallelism

In the previous two sections we have assumed that these various organs of indirect control *wanted* to work in the interests of the leadership and prevent plan and contract violations by ministerial officials. In several cases, however, the checking organs were organised along similar branch and functional lines to the ministries themselves. For example, both Gosplan's production department and the Party Secretariat's industrial departments were structured along parallel lines to the branch industrial ministries. Such parallelism had a tendency to breed common rather than conflicting interests. If a party and a ministerial official were both responsible for the performance of the coal industry, for example, they would find their overall task of raising production easier if both connived in illegal trading, hoarding, falsifying reports and so on rather than constantly quarrelling with each other in a bid to please higher authority.

Priority Sectors

Joseph Berliner has posed the question: 'How does one explain that in a totalitarian regime sturdily propped with all the paraphernalia of a police state, managers go blithely about hoarding materials, engaging in 'blat' and systematically evading the intent of regulations?'[61] An answer that has often been provided is that all this was in many cases acceptable to the leadership. The Soviet Politburo (so the argument runs) was not so stupid as to be ignorant of all the problems of the command economy. It knew that in spite of these problems it could enforce its basic priorities by keeping control over key sectors and expecting a certain limited compliance from other parts of the economy. Those in key sectors would behave themselves as they were under strict central control; those outside these sectors could ignore some regulations and plans provided always that they fulfilled the regime's basic requirements as indicated in periodic campaigns and the general aim of increasing physical output.

There is much evidence to support this view of Soviet economic policy-making after the war. Special emphasis was laid on certain priorities at certain times both by campaigns and new decrees and resolutions of the state and party leadership. Campaigns to direct attention to special problems have long been a feature of Soviet agricultural direction. They were also common enough in industry. For example problems in reconstructing the southern iron and steel industry in July 1947 prompted a conference of 'activists' in this sphere and editorials in the national press on this theme.[62] *Ad hoc* conferences attended by leading politicians (in this case Kaganovich, secretary of the Ukrainian party, and Tevosyan, Minister of Ferrous Metals) and editorials in *Pravda* and *Izvestiya* were a common way of inaugurating such campaigns. They were designed to focus the attention of officials and workers in the field on a new priority. For example, the 'Chutikh quality campaign' of February 1949 was a device for raising the priority of qualitative as against quantitative criteria in the textile industry.[63]

If the leadership's basic economic policy was set down in Five-Year Plans, its emphasis was also altered by interim decrees and resolutions. For example, the decree of the Council of Ministers of 23 December 1946 'On measures to speed up the development of state light industry producing mass consumption goods' altered plan targets and gave a higher priority to this sector of industry. The decree of 9 May 1950 'On lowering the cost of construction' was similarly designed to make officials in the construction industry pay more attention to efficiency rather than to the sheer volume of work done.[64]

These were the means by which the leadership communicated a change in their priorities in economic policy. To try to enforce these and the original plan's other priorities they had established arrangements for the direct administration of certain priority sectors. Some of these arrangements were more or less permanent; for example, central control over 'over-limit' capital projects and 'funded' production was clearly a device to ensure that basic economic priorities were enforced. Some factories were directly administered by the ministry (without intervening trusts or branch glavki);[65] these were presumably often the same priority plants where an inspector of the Central Committee Secretariat was permanently based. In addition several temporary bodies were established in the 1940s to strengthen central control over certain key sectors. For example, glavki for fuel supply and co-operative industry were established in 1946 and political departments on the railways in 1948.

The basis for the priority sector model is the argument that the command economy was a machine that could only achieve 'a limited number of well-defined objectives'.[66] In Chapters 4 and 6 we will examine two of the main objectives in post-war economic policy—the regional and the sectoral balance of the Soviet economy—to see whether even such aims as these could be achieved. Did the leadership realise the extent of the deviations in the command economy; were those deviations biased in particular directions so that, when aggregated, they might prevent the leadership achieving even its most basic priorities?

3 The Command Economy and the Formation of Regional Policy 1945–53

Introduction

One of the most important spheres of post-war economic policy for the Soviet leadership was concerned with priorities in the regional distribution of industrial growth. Some of Russia's earliest industries had based themselves in the eastern part of the country in the Urals and Siberia. Industrial development in nineteenth-century Russia was, however, almost entirely centred on the large cities of European Russia, notably Leningrad in the north-west, Moscow in the centre, and the Ukraine in the west. This pattern of regional growth continued largely unaltered throughout the first decade of Soviet industrialisation, with two notable exceptions. These were the building of the giant Urals-Kuznetsk iron and steel complex from 1930 and a decree of the following year severely limiting further industrial expansion within the confines of the over-crowded cities of Moscow and Leningrad. In spite of these the eastern areas of the USSR accounted for a smaller proportion of Soviet industrial output in 1940 than they had in 1917.[1] However, the regional policy issue really came to a head in 1939 when a firm policy favouring more development in the east was adopted in the Third Five-Year Plan. The Nazi occupation of much of European Russia during the war forced the leadership to concentrate even more industry east of the Ural mountains. The question facing the commanders of the Soviet economy of 1945 was whether to continue the eastern bias of the preceding six years or to concentrate their efforts on rebuilding the war-ravaged west of the Soviet Union.

Regional policy was not a simple matter of east against west, however. The criteria by which the regional balance of the economy was to be decided were several and complex and the regions involved by no means all 'eastern' or 'western'. These problems can clearly be seen in the formulation of regional policy in 1939 and during the war.

In the Third Five-Year Plan the Soviet leadership proclaimed its intention of promoting the 'complex development' of each economic region in the USSR.[2] Complex development meant the establishing of most basic branches of industry in each of the twenty regions. This meant a concentration of industrial growth on the less developed regions that lacked these basic branches. The less developed regions were mostly in the east of the Russian republic and in Central Asia. The map on p. 28 shows the economic regions used by planners in the 1940s. They differed from those of 1939 only in the addition of some new areas to the USSR in the war and the grouping together in economic regions of union republics that had previously each formed separate economic regions. The RSFSR remained divided into nine regions. Of these it was the Urals and the Western and Eastern Siberian and Far Eastern economic regions that were the least developed and the Central and North-Western regions the most industrialised. Of the other republics the most industrially developed was the Ukraine and the least the Central Asian republics.

The rationale of this policy of developing the USSR's more backward regions was found in five basic aims. These were the long-run maximisation of returns on capital investment, the reduction of administration and transport costs, the benefit of minority nationalities and the strengthening of Soviet defensive capability.

Perhaps the most important economic issue in regional policy is that of the time constraint. The Soviet leaders in 1939 decided that they could afford a longer time horizon than had previously been possible. For it was only in the longer term that investments in the less developed areas could be expected to yield greater benefits in the form of increased output and lower costs than investments in the developed regions of the west. The extra costs of investing in industrially underdeveloped regions can be summarised as follows:

1. Those of attracting workers with industrial experience to those backward areas *or* of transforming the local peasantry into a reliable industrial workforce.
2. Those of providing housing, drainage, roads, communications, power supplies, and all the other features of urban infrastructure that these sparsely populated areas lacked.
3. Those of transporting machinery and construction materials from the west, where they were produced, to build factories in the east.
4. Those of transporting the finished product to the main centres of consumption in the west.

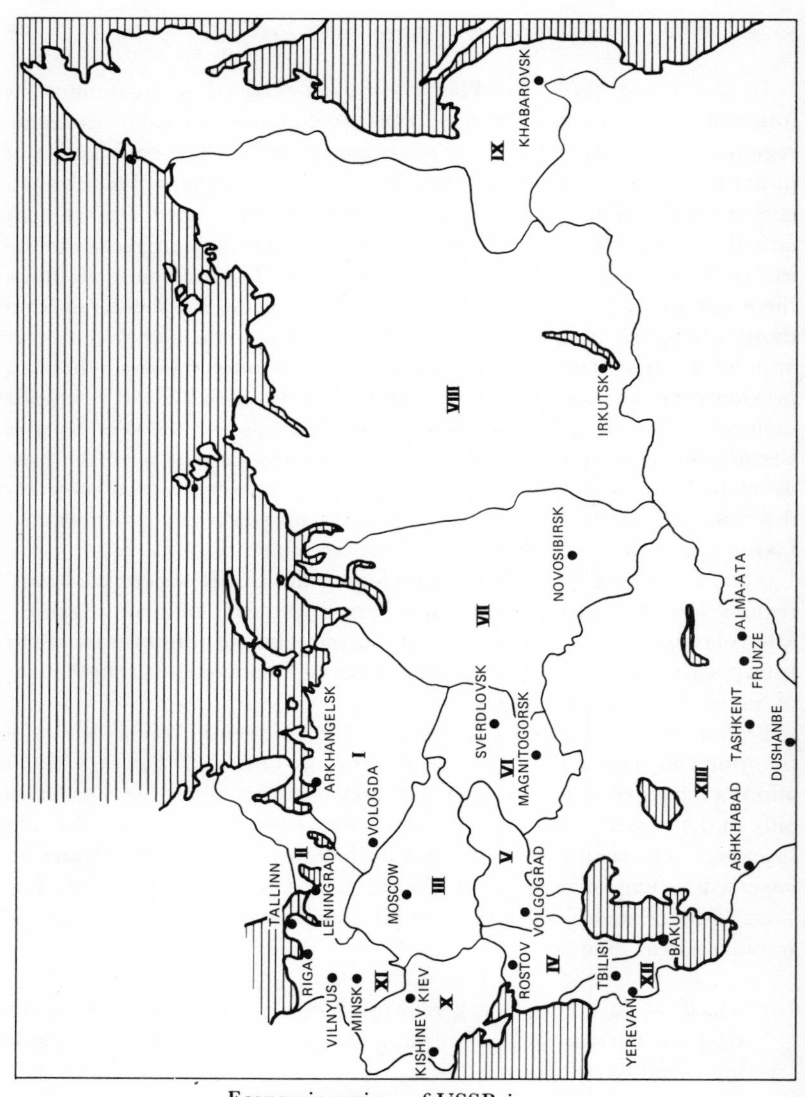

Economic regions of USSR in 1953

Key

RSFSR

I North

II North West

III Centre

IV North Caucasus

V Volga

VI Urals

VII West Siberia

VIII East Siberia

IX Far East

Other republics

X South (Ukraine, Moldavia)

XI West (Belorussia, Latvia, Lithuania, Estonia)

XII Trans-Caucasus (Georgia, Armenia, Azerbaidzhan)

XIII Central Asia and Kazukhstan (Kazakhstan, Uzbekistan, Tadzhikistan, Turkmenia, Kirgizia)

In other words, far more investment would be needed to yield the same returns in terms of finished product, and production costs would tend to be higher while those areas remained industrially underdeveloped.

These costs were reflected in the pattern of Russia's industrial development before 1917. In particular, the textile industry, using large quantities of relatively unskilled labour and selling its product in the mass markets of the European areas of the USSR, was heavily concentrated on the areas around Moscow and Petrograd. This was despite the fact that the raw materials for the industry (in the form of cotton and other crops) were grown far away in Soviet Central Asia. The major cost of investing in the west rather than the east was thus that of transporting raw materials. In the case of textiles this was outweighed by other costs. In the case of an industry like iron and steel, however, the costs of moving iron ore and coal in bulk were clearly a much greater constituent of total production costs. Even in this case the relative proximity of the Don coal basin and the iron ore deposits at Krivoi Rog to large centres of population and industry in the Ukraine (and even the Central region) made them a far more attractive proposition to investors than the far distant Kuznetsk basin (in Western Siberia) and the iron ore of the Urals region. In spite of much of the raw material reserves for the iron and steel industry being located in the eastern areas (on modern estimates more than 90 per cent of the USSR's potential energy sources and 40 per cent of its known reserves of iron ore)[3] the Old West still accounted for more than 60 per cent of the output of Russia's extractive and processing industries in 1912.[4] Only in the longer run could the high initial costs of construction and production in the east be outweighed by economies resulting from the ready availability of raw materials in these areas.

Soviet economists were (and are) not slow to argue that the drive for short-run returns is characteristic of the capitalist mode of development. Whilst capitalists seek rapid returns on investments and so tend to invest in developed regions,[5] a socialist state can afford to forego profits to the individual enterprise in the shorter run with an eye, firstly, to yielding profits to a wider spectrum of society (by building up the social infrastructure of underdeveloped regions) and, secondly, discounting those initial losses against longer term gains in the form of lower production costs resulting from the readier availability of raw materials. In terms of economic theory, greater weight is placed on social rather than private returns and a much lower rate of time preference is exhibited.

These factors began to have an impact on the minds of Soviet planners as early as 1921 with the plan for the electrification of the whole of Russia (GOELRO). It was only with the announcement of the Third Five-Year

Plan, however, that any concerted effort was made to enforce a policy that sought to exploit the long-run potential of eastern areas. Until then the scarcity of capital resources had dictated a much greater time constraint.

In general a regional policy based on short-run maximisation of returns on investments favoured the western areas of the USSR and the sort of longer-run aims pronounced in 1939 favoured the less industrialised east of the country. Within these broad categories of east and west, however, there were many variations in levels of industrial development. The quickest returns could be expected, other things being equal, from the old industrial areas of the West, the Central, North-Western and Southern economic regions. These areas, that were the focus of industrial development until 1939, we shall term the 'Old West'. Other areas of European Russia such as the Northern Caucasian economic regions had still seen relatively little industrialisation. The complex development aims of the Third Five-Year Plan did favour them. Similarly, not all of the eastern regions were completely lacking in industrial might. The Urals, Western Siberia and parts of the Central Asian and Kazakhstan region had seen some industrial growth by 1940. It was in these 'Near Eastern' regions that investments would yield returns within the foreseeable future. The Eastern Siberian and Far Eastern economic regions had very little industry at all and it would be many years before a new plant there could begin to produce good quality output at reasonable cost. As a result the debate over time horizons amongst regional policy-makers was often couched in terms of (in effect) Old West and Near East rather than simply west and east. In addition there were areas that some classified as eastern and some as western, notably the Volga and TransCaucasian regions, both of which had seen a degree of industrial development by 1940.

The policy of complex development was not based solely on the desire to lengthen time horizons to enable the exploitation of rich raw material deposits in the east. Administrative and transport costs also weighed heavily, as did more political factors like defence and nationalities consideration.

The continuing growth of the Old West during the first decade of Stalinist industrialisation had its costs in terms of administrative complexity. Construction projects and enterprises in production in these areas were becoming so large by the early 1930s that it became difficult to co-ordinate all their activities and especially to secure regular supplies of inputs on the required scale. Furthermore, the infrastructural facilities (fuel and power supplies, housing, and so on) of the large cities in these areas were being overstretched. As a result construction projects were taking far too long to complete (if they were ever finished) and factories suffered numerous dislocations in supplies of inputs. In response to this problem of *gigantomania* the Soviet leadership placed severe restrictions on further industrial expansion in Moscow and Leningrad in 1931 and in five other cities in 1939.[6] The threat of administrative diseconomies of scale thus tended to turn planners against further development of the Old West.

The aim of complex development was also to cut down on the transport of goods between regions. Each region was supposed to become increasingly self-sufficient. The transport sector was a major bottleneck in the Soviet economy throughout the 1930s. The volume of railway traffic grew 2.2 times faster than GNP over the period 1928–40. The only ways out of this bottleneck in the long term were to invest in new transport facilities or to cut down use of the railways by industry. The Soviet government preferred the latter course. In the Third Five-Year Plan it called for new plants to be built nearer to their sources of raw materials (and to their major markets, although this criterion was often left to one side). Existing plants were also to make more use of locally produced raw materials and power sources. To reduce transport costs, therefore, new industrial projects were increasingly to be sited near to the abundant raw materials of the eastern areas.

The regional policy of 1939 was not solely dictated by economic costs and benefits. It also envisaged some benefits to its policy on the minority nationalities. The industrial development of such backward areas as Soviet Central Asia would surely promote greater loyalty to the Bolshevik government on the part of these non-Russian peoples. Industrialisation would break down traditional loyalties to 'bourgeois nationalist' leaders and establish a new proletarian class which would identify far more strongly with the 'ruling Russian proletariat'. Nationalities policy was a feature of Soviet regional policy from the GOELRO plan right up to the Third Five-Year Plan with its commitment to the further economic and cultural development of national republics and *oblasti* (provinces). Such a criterion might favour some western areas such as the Ukraine and Belorussia rather than the eastern regions of the RSFSR which were sparsely populated and had a large proportion of ethnic Russians. A new twist to the nationalities policy was provided by the addition of many new areas to the USSR in 1940. Clearly investments here might buy a degree of political loyalty. Areas such as the Baltic States were generally more developed anyway than the regions annexed from Poland and added to the Ukrainian and Belorussian republics of the USSR. Nationalities considerations did not always dictate investment in the east, and it could cut across the long-run maximisation aim (as in the case of the Baltic States).

The Soviet Union's defensive capability also dictated further industrial development of the east. The decision to build the Urals-Kuznetsk combine in 1930 was not based purely on long-run maximisation considerations. It also reflected the desire to build up heavy industrial plants in the eastern areas far distant from any front line in the event of a war. The distribution of Soviet industry in 1930 meant that any determined invasion from the west could result in the USSR losing almost all of the heavy industrial base for its armaments industry. It was only because of such decisions as these of 1930 and 1939 (when it was decided to establish 'duplicate' heavy industrial enterprises in 'hotbeds' (*ochagi*) of industrial growth in the eastern areas and the Volga region) that the Soviet Union was in any position to recover from the Nazi onslaught of 1941. Considerations of national defence therefore also dictated a transfer of developmental resources to the Near East. They may, however, have militated against industrial investment in areas like Eastern Siberia and the Far East that could prove too far from the front line and in areas like the TransCaucasus that could be (and in effect were in 1942) cut off from the centre of Russia by a determined invader.

The fulfilment of the regional policy of the Third Five-Year Plan was cut short in June 1941 by Hitler's invasion of the Soviet Union. No dramatic changes in the regional distribution of output could have been expected in only two years. Some progress was made, however, in carrying out the new policy. The pressing nature of the transport problem dictated changes in railway freight tariffs in April 1939 that effectively discouraged many over-long freight hauls; their average length fell from 718 to 700 km. between 1938 and 1940 in spite of a steady increase in that figure before 1938.[7] The first annual plans for the new economic regions were drawn up in 1940. Longer-term perspective plans for the development of these areas were to be part of the Fifteen-Year Plan project that was to go into operation in 1943. This project was, of course, never completed. It is, however, noteworthy that at this stage some thought was apparently given to the setting up of separate administrative organs in each region to promote the aim of complex development. At the instigation of its new head, N. A. Voznesenskii, Gosplan set up specialised *apparats* of thirty to forty people in each economic region, each under a Gosplan 'representative' (*Upolnomochennyi*). Each of these was in turn subordinate to the nine special departments for territorial planning established in Gosplan's Moscow headquarters. The main task of these local and central bodies was, according to one representative, 'to . . . systematically check on the execution of the state plan, to take operative measures to check on and secure the complex development of each economic region of the country'.[8] Nevertheless the government organs of Union republics still lacked any formal authority over large-scale industry (of All-Union significance), and in none of the economic regions of the RSFSR did a unified government or party organ even exist. The implementation of regional policy decisions was of necessity left to the All-Union production-branch ministries in Moscow. It was really the necessities of war that forced them to accede to this policy.

The Impact of the Second World War on Soviet Regional Policy

The Nazi invasion of the USSR began in June 1941. Within six months the German forces had occupied or isolated most of the large industrial centres west of Moscow; by the end of 1942 they held areas that had produced 30 per cent of Soviet industrial output before the war, and still threatened others. In order to make good these losses the Soviet government decided to force the growth of industry in the Near East (including the Volga region) at a rate barely dreamt of before the war. As a result, as a Soviet economist wrote (in the last months of the war itself):

> The changes that have taken place in the economy of the mid-Volga, the Urals, Western Siberia, Kazakhstan and Uzbekistan were in complete accord with the directives of the 18th Congress of the VKP(b) on the creation of duplicate enterprises in the eastern regions, on the development in them, by all measures, of metal industries and of coal mining, on the creation of new large bases of the textile industry, on the drawing closer of industries to the sources of their raw materials and areas of consumption, and on the complex development of basic economic regions.[9]

The changes in the actual distribution of industry during the war were the result of, first, the occupation and destruction of enterprises in the western areas; second, of the evacuation of machinery and men from west to east in 1941–2; third, the rapid expansion of capital construction in areas east of Moscow in 1941–3; and finally, the reconstruction during the war itself of enterprises in areas liberated from Nazi occupation over the period 1943–5.

According to the Soviet Committee on Reparations, economic assets valued at 679 billion roubles were destroyed during the war. Khrushchev later talked of human losses of more than 20 million. The greater part of this burden fell on the occupied areas of the west; two-thirds of their 'national property' was destroyed.[10] In the Ukraine, Belorussia and the occupied areas of the RSFSR less than 20 per cent of the industrial labour force survived the occupation (some were lucky enough to be evacuated to the east, some were called up, many died in German hands in Poland and elsewhere); and more than 80 per cent of industrial enterprises in these republics were put out of action by the occupation and subsequent liberation of these areas. In total some 31,850 large industrial enterprises were put out of production during the war.[11]

In contrast only 1700 or so such enterprises were lost to western industry through evacuation to the east over the period 1941–3.[12] It was impossible to evacuate anything but relatively light and easily dismantled machinery and, of course, workers (about one-third of the workforce were evacuated with their enterprises).[13] Heavy plant, extractive enterprises, buildings and the like had to be left to the invader or (preferably) destroyed to prevent their exploitation by the Germans. Limitations of transport facilities and time (the German advance was so rapid as to take them to the gates of Moscow in five months) also prevented the evacuation of such machinery from the areas further west; far more enterprises were evacuated from Moscow oblast than from the whole of Belorussia, and, according to some estimates, from the whole of the Ukraine.[14]

The losses to the economy of the west from evacuation were therefore of little significance compared to its direct losses from destruction. However, the equipment and workers evacuated to the east were vital factors in the expansion of eastern industry in 1941–2. Evacuated machinery could be re-established in the east far quicker than new plant could be built from scratch in the same area. According to one estimate some 47 per cent of the equipment functioning in factories in the eastern areas (including the Volga region) in 1942 had been evacuated from the west.[15] By the spring of 1942 5.9 million Soviet citizens had also been evacuated to the eastern areas of the RSFSR,[16] the pre-war population of which was 43.2 million.

New construction in the Near East was also proceeding apace in the first two years of the war. The proportion of Soviet capital investment in industry directed to the Urals and Western Siberia rose from 13.1 per cent in 1940 to 38.7 per cent in 1942.[17] As a result of both evacuation and recruitment into industry of local *kolkhozniki* (collective farm workers), the number of industrial workers in the Urals and Volga regions rose by some 65 per cent between 1940 and 1943.[18] The result of evacuation, new construction and labour recruitment was that the industrial output of the Near East rose 2.9 times between 1940 and 1943, whilst that of the USSR as a whole fell by 10 per cent.[19] Clearly there was a massive shift of the focus of Soviet industry to the Near East (probably far greater than ever envisaged in the Third Five-Year Plan) during the war.

However, the extent of that shift and its implications for post-war policy should not be overestimated. In the first place, most re-established and new constructed plant was concentrated on the more developed areas of the eastern regions, such as the Volga, the Urals, and West Siberian regions and the Uzbek and Kazakh republics. These areas were not only better equipped in terms of existing industrial and urban facilities to accommodate new industry than were Eastern Siberia and the Far East and the smaller republics of Central Asia, but were mostly nearer to the armies in the west that consumed much of their output. Indeed the August 1941 decree that signalled the drive for the development of eastern industry did not refer to East Siberia or the Far East. Of the 1523 evacuated enterprises only 78 were re-established in East Siberia, none in the Far East, and fewer than 70 in the smaller Central Asian republics.[20] The share of East Siberia and the Far East in the (declining) total of capital construction in the USSR actually *fell* between 1940 and 1942 and again in 1943.[21] The industrial output of the East Siberian and Far East economic regions rose by 25 per cent and 15 per cent respectively over the period 1940–5; that of Tadzhikistan and Turkemenia *fell* by 26 per cent and 13 per cent respectively. Not all the eastern regions benefited equally from the redistribution of industry during the war.

Second, the data presented above may exaggerate the extent of wartime industrial growth in the east in so far as much of that growth was concentrated in the defence sector. Of the 1523 enterprises evacuated eastwards in 1941, 1360 were part of the 'war industry'. The Urals alone produced 40 per cent of the output of that branch of Soviet industry.[22] As the economic historian Kravchenko has noted, the multiplier effects on the region's economy of defence production are less than those of civilian output; the former produces 'means of destruction' and the latter 'means of construction'.[23] This was especially true of the eastern enterprises of the war commissariats, most of whose final output went to the armies in the west. There was almost certainly a large 'export leakage' in the reproductive process of the economy of the eastern regions. The leakage of real products exported westwards was not matched by financial inflows to the east that could be used to finance future development. The lack of financial discipline in the Soviet economy meant that goods were not always paid for very promptly, and, in any case, finance for investment came almost entirely from central funds rather than the enterprises' own profits (many enterprises could not make profits under the existing price structure). However, the new plants built in the Urals or Siberia to produce arms, munitions, and uniforms did add to the productive capacity of the Near East; and their reproductive capacity could be increased after the war when they were converted to civilian production. These new investments in the east also added to the pool of skilled labour and to the fund of wages available for consumption in these areas. However, the costs of reconversion to civilian output proved sufficient in 1945 and 1946 to offset a sizeable portion of the wartime gains of the economy of the eastern regions. The industrial output of Siberia and the Far East fell by some 20 per cent in 1946.[24] In addition the high priority accorded to defence industry meant that other sectors in the east often suffered. The output of foodstuffs, timber, and even construction materials in Siberia and the Far East *fell* substantially during the war.[25] To claim that the war gave the eastern areas a greatly expanded base for post-war industrial growth is to ignore the disproportionalities between sectors within that base. The light industries of the east as well as heavy and light industry in the south had to be reconstructed after the war.[26]

In fact the very rate of expansion in the defence industry was bound to lead to disproportionalities, and so to bottlenecks and problems in exploiting new and evacuated plant. Figures for evacuations and capital construction alone may exaggerate the growth of eastern industry in the first half of the war. Often the machinery could not be used because of shortages of such basic resources as power and construction materials. The poor housing and food situation must also have limited the effectiveness with which new capital resources were used in the east.[27] Some of the east's wartime gains in output levels were also due to exploitation of previously unused capacity and the institution of compulsory overtime:[28] both these were once-and-for-all increases, the latter of which was to disappear in peacetime.

The urgency of wartime also meant that peacetime norms for construction methods, exploitation of machinery, housing, and so on, were violated, shortening the life of both buildings and equipment.[29] The quality of wartime construction in the east was thus significantly lower than that of pre- and post-war construction in the west.

Individual construction projects were also conducted on rather an *ad hoc* and unco-ordinated basis. The August 1941 plan for the development of the Near Eastern areas in 1941–2 was really little more than a hurriedly drawn-up statement of pressing priorities.[30] Only in 1943 was the first long-term perspective plan for an eastern region (the Urals) drawn up; and by that date the rate of new construction in these areas had already begun to tail off.

The reconstruction of the war-ravaged industry of the west began long before the end of the war, as early as in 1942 in the Moscow coal basin. Large-scale capital reconstruction, however, dates from the August 1943 decree of the Central Committee and Sovnarkom 'On urgent measures for the reconstruction of the economy in the areas liberated from German occupation'.[31] Thereafter resources were concentrated on the reconstruction of western industry rather than on capital construction in the east. In fact the absolute level of capital construction in the eastern regions (including the Volga) fell between 1943 and 1944; in consequence only 36.7 per cent of USSR capital investment in 1944 went to the Urals, Siberia, the Far East and Central Asia and Kazakhstan, compared to 61.5 per cent in the previous year.[32] A number of construction teams were in fact returned to the west as they could find no employment in the east.[33] The eastern bias of regional policy in the first half of the war was at least partially reversed during the last two years of the conflict.

This reversal was also reflected in the movement of the labour force. Whilst re-evacuation of plant was limited to unutilised equipment (except in some sectors of light industry),[34] '1944 saw a vast amount of organizational work on the redistribution of the qualified labour force . . . between the economic regions of the nation, especially between the eastern and western regions'.[35]

In addition the reproductive effect of eastern industry in the period 1943–5 was curtailed by a similar export leakage to that operating in the first two years of the war. Now much of the output of civilian industry in the east was sent to the west to help in the reconstruction effort. Provinces and factories in the east were organised to send aid (*shefstvovat'*) to the stricken areas of the west. The overall impact of this aid can only be guessed at; however, one modern student of the reconstruction process did not hesitate to describe these events of 1943–4 as a significant 're-distribution of resources' in favour of the west. He also noted the much better fulfilment of reconstruction than of construction plans in the ferrous metal industry at this time.[36]

By 1944 the 'liberated areas' were consuming 41.6 per cent of the USSR's total investments, compared to only 16.3 per cent in the previous year: *Narkomstroi* was doing 42 per cent of its work in the liberated areas in 1944, compared only to 1 per cent in 1943 and only 29 per cent even in 1940.[37] Such was the extent of the drive for reconstruction in 1944 that the Ukraine was taking a greater proportion of Soviet capital investment expenditure in that year than it had before the war!

In addition, reconstruction from 1943 was conducted in a more organised fashion, and in accordance with longer-run peacetime perspectives, than new construction in the east had been in the earlier war years.

First, reconstruction at least from 1944 was concentrated on the civilian sector of heavy industry, rather than the armaments and munitions sector that was so important in the earlier expansion of the eastern regions.[38] Losses due to non-reproductive output and post-war reconversion problems were not thus as important in the Ukraine and Belorussia as in the Urals and Siberia. Most of the output of reconstructed plants was also retained in the western areas, avoiding the east's export leakage problem.

In addition, destroyed plants were often reconstructed using the latest machinery. In contrast, as we have noted, many of the new plants in the east were based on old evacuated machinery very little of which was subsequently re-evacuated. Indeed destroyed plants were sometimes rebuilt almost from scratch, using fewer of the corner-cutting construction techniques earlier employed in the east. Increasingly reconstruction became the business of specialist (*podryadnye*) construction organisations rather than bands of less than skilled volunteers.[39]

The more permanent nature of many individual reconstruction projects was accompanied by a greater effort to combine these projects in long-term perspective plans. As we have seen, the first perspective plan project for an eastern region was announced only after capital allocations in that area had begun to fall. The first draft of a plan for the liberated areas (for 1943–7) actually appeared in 1944, at the height of the reconstruction process. This project was presumably the work of the new Gosplan administration for reconstruction, established in February 1943, and merged with the administration of regional planning in June 1944.[40] The construction process in the east had by contrast been less co-ordinated, although there had existed an *ad hoc* commission, headed by N. A. Voznesenskii, to sort out bottlenecks in the exploitation of evacuated plant in 1942. (A similar 'Committee for the Reconstruction of the Liberated Areas' was established in the spring of 1943.)

Finally, it must be emphasised that to compare the relative economic health of western and eastern areas at the end of the war solely in terms of gross output figures is misleading. The industrial output of the liberated areas in 1945 was only 30 per cent of its 1940 level.[41] However, by mid-1945 some two-thirds of the industrial enterprises in those areas were back in production.[42] To get an enterprise back into production demanded a greater expenditure of both capital and labour resources than merely to increase the output of an already functioning enterprise (that is, one already 'back into production') by a similar amount, due to the costs of repairing fixed capital like buildings and power supply networks. Given the proportion of enterprises back in operation by mid-1945, therefore, one would expect a given capital expenditure in the liberated areas to yield a much greater increase in output in 1946 than in 1945. That the base for further reconstruction in the west was much stronger than output figures alone would suggest is also reflected in the fact that the plants producing machinery for the coal industry of the South region had regained 83.7 per cent of their 1940 capacity by 1 July 1945, thus providing a healthy base for reconstructing the mines themselves.[43]

Given these qualifications to the use of gross output data, the overall impact of wartime destruction, evacuation, construction and reconstruction is summarised in Table 3.1. It is clear from this table that the effect of the war on particular economic regions was no more uniform in the west than it was in the east. Of the occupied areas, Belorussia and the Ukraine were the worst hit although their new western oblasti and the Donbass were high on the list of priorities in the reconstruction effort. In spite of the high cost of rebuilding the Don coal basin (including the pumping of millions of gallons of water from its shafts), it had regained 40 per cent of its pre-war output level by 1945.[44]

The new areas that had been added to the USSR in 1940 had to be shorn of 'capitalist' survivals. The drive to 'build socialism' in (and in particular, to speed up the industrial development of) the western areas of the Ukraine, Belorussia and Moldavia and of the three Baltic republics meant that these areas were among the highest priorities in the reconstruction effort. Indicative of this was the establishment of the special *buro* for the affairs of the Baltic States and Moldavia that was directly subordinated to the All-Union Central Committee and thus to the Politburbo and Secretariat.[46] Nationalities considerations clearly influenced regional policy-formation during the wartime reconstruction process. As a result these new areas were in much healthier economic position in early 1946 than the old parts of the Ukraine and Belorussia, in spite of being amongst the last areas of the USSR to be liberated.[47]

In the RSFSR only the North-Western and North Caucasus regions, both centres of heavy and long-drawn-out fighting during the war, suffered especially badly. In contrast, the Northern and Central economic regions suffered only a small diminution of industrial wealth during the war. Both were only partially occupied and were the centres of heavy construction work (notably in the Pechora and Moscow coal basins) as well as destruction and early reconstruction during the war. Parts of the Volga region around Stalingrad were also occupied for a time but at the same time massive industrial construction (especially in oil and related industries) was taking place to the north of that city, with the result that the region as a whole gained substantially in economic importance during the war. The Volga cannot be classified simply as an eastern or western area or as an occupied or rear zone. The only remaining area of the Soviet Union,

TABLE 3.1. The impact of the Second World War on the distribution of industrial output in the USSR

	Industrial output in 1945 (at 1940 = 100)	Share of USSR industrial output (%) 1940	Share of USSR industrial output (%) 1945
USSR	92	100	100
of which:			
liberated areas	30	33	10.8
RSFSR	106	72.6	83.6
of which:			
Economic regions: Centre	82	57.6 (Centre, North, North-West, North-Caucasus combined)	44.6 (Centre, North, North-West, North-Caucasus combined)
North	92		
North-West	33		
North-Caucasus	41		
Trans-Volga	181	2.8	5.5
Urals	305	6.6	21.9
West Siberia	270	2.7	7.9
East Siberia	128	1.4	1.9
Far East	112	1.5	1.8
UKRAINE	26	17	4.8
of which:			
'New' Western oblasti	35		
LITHUANIA	40	0.29	0.19
LATVIA	47	n.a.	n.a.
ESTONIA	73	n.a.	n.a.
GEORGIA	80	1.0	0.87
ARMENIA	93	0.31	0.31
AZERBAIDZHAN	78	2.1	1.8
UZBEKISTAN	107	1.6	1.9
KIRGIZIA	122	0.20	0.27
TADZHIKISTAN	74	0.23	0.19
TURKMENISTAN	87	0.31	0.29
KAZAKHSTAN	137	1.1	1.6
BELORUSSIA	20	1.7	0.37
MOLDAVIA	44	0.19	0.09

Source: See Note 45 on p. 157.

that of the Caucasian republics, did not suffer invasion, but was partially isolated from most of the other industrial centres of the USSR for most of the war. Whilst suffering few direct losses, its industries did tend to stagnate; for example, Azerbaidzhan's share of Soviet oil output fell markedly over the war years.

The post-war issues in regional policy were not therefore purely ones of Near East–Old West balance. However, this balance between new and old industrial areas remained of basic importance. One student of the war economy put the issue very clearly when he wrote in 1947:

> But these changes in the distribution of industry are not temporary or transitory. In the post-war period, when the level of industrial production of 1940 in the western areas will be regained and even exceeded, the pre-war relationship of the regions in the industry of the USSR will not be restored. The old industrial regions will not regain their industrial superiority, as the eastern regions will continue their industrial development.[48]

Should the post-war economic effort be directed solely to the total reconstruction of the western areas, or should wartime gains of the Near Eastern areas be consolidated at the (partial) expense of the reconstruction effort? This was the major issue facing regional policy-makers in 1945. This issue could be resolved into that of whether to maximise short-run or long-run returns on investment. Such subsidiary issues as the rates of growth to be sought in East Siberia could be discussed in similar terms. However, the priority to be accorded to the development of the new areas, the North and TransCaucasus, and the Central Asian republics also depended on nationalities policy considerations. Extensive wartime damage to the railway network also put the transport factor firmly in the minds of policy-makers in 1945. With a victorious army occupying half of Europe, the Soviet Union after the war was surely less mindful of national defence considerations than in previous years. Finally, in the general chaos of the post-war reconstruction of industry, administrative problems were so great that they must have impinged on regional policy-making, although in 1945 their full implications for that policy were not at all clear.

Before discussing the debates over post-war regional policy, it must be noted that the administrative structure of the command economy during the war years did resemble in many respects the priority sector model. The GOKO kept a very close check on the implementation of its decisions in certain key spheres. This was partly done by sending *upolnomochennye* to the localities to investigate the situation and report back to the centre. In addition Gosplan and the ministries were obliged to report very frequently to the GOKO (in some cases twice a week) on the progress of work in priority areas. The Committee also created several *ad hoc* commissions to co-ordinate and supervise the execution of the most important facets of policy. For example, those charged with the supervision of the evacuation of 1941–2 and the reconstruction of 1943–5 were both headed by Voznesenskii; another committee was established to recruit labour and direct it to key projects.

Such methods of central control could only be used in a limited number of policy areas; only a certain amount of information could be relayed to one such small body at any one time. Central control over the major priorities—first, the defence industries and, latterly, the most important reconstruction projects (for example, the Donbass)—seemed fairly strict. As a corollary, of course, the less important areas (for example, the small Central Asian republics and the Far East) and sectors (especially light industry) were often left to fend for themselves. They ranked very low in the priorities for the allocation of centrally funded materials, but were relatively free to organise themselves and use what local resources they could. For example, the author has been assured that much of the reconstruction effort in Belorussia during the war was carried out at that republic's own cost in materials and labour. To what extent the centre could enforce a co-ordinated regional policy through such a framework, or whether the administrative system should be amended to facilitate that enforcement, would prove an issue in post-war regional policy debates.

Regional Policy Discussions before the Fourth Five-Year Plan

The first major statement on post-war economic policy was to be the new Five-Year Plan, approved by the Supreme Soviet in March 1946. Officially work had begun on the plan document only seven months previously (three days after the Japanese surrender), but the policies to be pursued in the plan had been under active debate as early as 1943. The regional policy embodied in the new plan was significantly influenced by these debates amongst planners, administrators, academics and leading political figures, particularly during 1945. Both staff and line officials of the command economy sought to influence the leadership's plans for the regional distribution of industrial growth.

The two main schools of thought on post-war regional policy were clearly expressed in discussions within Gosplan in 1945. The immediate problem was that 'regionalisation' (*raionirovanie*); that is, the nature of the criteria according to which the USSR should be split into economic regions for planning purposes. On 5 July 1945 the chairman of Gosplan (presumably N. A. Voznesenskii)[49] presented the findings of an investigatory committee to a session of the State Planning Commission. In his report Voznesenskii—a candidate member of the Politburo and a former GOKO member—proposed that the USSR be divided for planning purposes into seventeen regions, each distinguished by its current specialisation in the output of a particular branch (or branches) of industry.[50] Such a regionalisation would have inhibited the planning of development of industries completely new to a particular area. It would therefore have discriminated against the more backward areas of the east, where relatively few branches of industry were well developed at this time. Voznesenskii (or at least the officials who had prepared the report) were favouring the maximisation of shorter- rather than longer-run returns.

This regionalisation would also have accorded a very low priority to nationalities policy. The divisions between regions often ignored national boundaries; for example, the Smolensk oblast' of the RSFSR was included in the same (projected) region as the Belorussian republic. If this regionalisation were to form the basis for future regional planning, how could the balance of industrial development amongst the national regions of the USSR form any part of regional policy?

In opposition to this report A. V. Korobov, head of Gosplan's Capital Construction Department, called for a regionalisation based on the need to ensure the complex development of each region. In addition he stressed that the regionalisation should take account of the changes in the pattern of industrial location wrought by the war. In saying this Korobov was reiterating his 1941 advocacy of a policy favouring the eastern regions.[51] In that year he had argued for the complex development of each region on the grounds of reducing transport costs, of improving industrial administration (by ensuring better co-ordination of different enterprises within the same region), of benefiting the more backward national groups, and of maximising long-run returns on capital. The only argument for complex and eastern-biased development that Korobov did not use (in 1941) was that of strengthening defence capabilities! It seems reasonable to assume that Korobov articulated similar views in 1945. If so, he was proposing a set of regions, each of which could be made largely self-sufficient in the output of most of the major branches of the national economy—fuel, machinery metals, construction materials, consumer goods and foodstuffs, and agricultural produce.

It was Korobov's views that prevailed rather than those of his exalted superior Voznesenskii. On 25 July 1945 a new commission was established to work out a regionalisation based on complex development criteria. Within Gosplan at least, policy was not always handed down from above but could be debated by interested parties with some prospects of affecting official policy. As we shall see later the main obstacle to the advocates of complex development was not their superiors but rather their fellow officials within branch ministries and even within Gosplan itself. In this particular case the new commission's report was never made public and the Five-Year Plan utilised the old twenty-area regionalisation.

Besides Gosplan's officials many other staff advisers and line officials within the structure of the command economy voiced their views on regional policy. The specialised small circulation journals published by state and party organs, as well as meetings such as those of the Supreme Soviet, provided a public glimpse of the debates that were going on in private. For example, the economist E. Granovskii strongly argued the case for complex development in the columns of *Bol'shevik* (the Party's theoretical journal, not normally so concerned with such practical economic issues).[52] He produced all the familiar arguments in favour of complex development and the eastern regions, citing the five criteria of the Third Five-Year Plan. Granovskii went further than most other commentators in calling for each region to be largely self-sufficient in most forms of industrial output and in transport facilities, as well as in the more usual spheres advocated earlier by people like Korobov.

The argument of Granovskii and Korobov that the successes of wartime development in the east showed its potential for the future was frequently heard at this time. It was advanced, for example, by A. Lavrishchev (later a member of the second Gosplan commission mentioned above) in *Planovoe Khozyaistvo* (Gosplan's own organ) and by I. Lerskii in a pamphlet also published by Gosplan.[53]

By no means all planners took up the cause of the eastern regions: many were concerned to stress the potential of the partially reconstructed western regions. The well-known economic commentator B. M. Sukharevskii emphasised the advantages of reconstructing western industry on the basis of the latest technology. A. Baikov and A. Zelenovskii both made much the same point in the pages of *Planovoe Khozyaistvo* in January 1946.[54]

Such debates as these were aimed at influencing leadership policy through Gosplan and the other organs involved. It was to planners and academics that the leadership looked for advice in such matters. Line officials also provided information that could influence the leadership but it was often of a more empirical and detailed nature. For example, at the Supreme Soviet session of April 1945 a number of deputies from the liberated areas of the west (many of them party officials) attacked what they saw as an overconcentration of resources on the areas east of the Urals.[55] Where planners and academics tended to look to the future, line officials had an eye on current practice and its defects.

It is one thing to demonstrate the existence of debate on an issue and quite another to show that it influenced leadership decisions. To illustrate this we must show how the leadership itself was involved in regional policy and how it spelt out its decisions in the new Five-Year Plan.

Several leading Soviet politicians publicly hinted at their views on leadership policy, although they could not be as forthright as less senior figures. For example, in an election speech at Gorki in February 1946 Voznesenskii emphasised the development potential of the liberated areas of the west and made no reference to either the wartime successes or the future potential of eastern regions.[56] In contrast Zhdanov, speaking in Leningrad a few days later, emphasised that further development of the east was 'absolutely advisable and necessary'.[57] It was his supposed party revivalist ally Molotov who stressed rather a different feature of regional policy, the nationalities factor. On 7 November 1945 he emphasised the need to direct resources towards the new areas of the west of the USSR on grounds of benefit to those national areas rather than on the basis of economic returns to the Soviet Union as a whole. Oddly enough the journal *Bol'shevik* omitted all reference to this point in its commentary on Molotov's speech.[58] Perhaps some of its editorial board could not agree with him. A failure to comment on a view was often a covert way of criticising it in the Kremlinological climate of the times.

Divisions within the leadership partially reflected divisions of opinion amongst more humble officials. The decisions eventually reached and announced in March 1946 were essentially a compromise between advocates of east and west, with the former perhaps coming out on top. The principle of complex development was reaffirmed but not to quite the extremes advocated by Granovskii or Korobov.

Regional Policy Decisions 1946–7

Before proceeding to a detailed examination of the provisions of the Fourth Five-Year Plan (1946–50) on regional policy we must consider the argument that the directives of Five-Year Plans were never intended to be concrete guides to official policy.[59] The Fourth Five-Year Plan was apparently produced in a very short time. The plan formation process had officially begun with a decree of 19 August 1945 and the plan was approved by the Supreme Soviet only seven months later. That could hardly have allowed sufficient time for consultation of all the relevant sources of information and advice, especially in the administrative and economic chaos that followed the war (resulting from wartime destruction and the reconversion of the economy to civilian output). In addition the plan was announced in March 1946, some three months after it was supposed to be operative (this delay has occurred with all the Soviet Five-Year Plans to date). Concrete economic direction (it is argued) was embodied in annual and quarterly plans, the longer-term targets amounting to little more than window-dressing.

Such Western analyses are in stark contrast to the attitude prevailing amongst Soviet economists after the war. They were almost unanimous in stressing the virtues of 'perspective planning' over longer and longer periods.[60] It is surely more accurate to interpret such perspective plans as no more than general guides to policy, a statement of overall priorities and an assessment of their implications for various sectors and regions, and not as a detailed set of directives for the day-to-day running of the economy. As such, they were indeed definitive statements of the leadership's priorities in economic policy.

Neither were Five-Year Plans quite such vague descriptions of future aspirations as some authors would have us believe. The Fourth Five-Year Plan directives were in fact the result of many months of detailed preparatory work and consultations. Work on the plan did not suddenly start with the decree of the Council of Ministers and the Central Committee of 19 August 1945; a great deal of expert advice and information had been sought within Gosplan and other agencies long before that date, and certainly from at least 1943.[61] As we have seen, the substantive issues had been under discussion long before August 1945.

The directives drawn up by the planners and approved by the leadership were far more precise than the published document itself reveals. For example, Voznesenskii, in his speech presenting the draft plan to the Supreme Soviet, gave a more detailed breakdown of investment plans, and Gosplan's journal published a series of articles filling in extra details of the plan as it concerned particular sectors of industry and individual regions.[62]

In its approved form the plan document did not make such clear-cut reference to regional policy as had its predecessor of 1939. However the plan, unlike both the Third and Fifth Plans, did devote a separate section to each of the Union republics. In presenting the draft plan to the Supreme Soviet, Voznesenskii clearly reiterated the policy of complex development approved by the 18th Party Congress.[63] In a bid to eliminate overlong and irrational transports of products each economic region of the USSR was to develop its own fuel and power base, and to become largely self-sufficient in the output of construction materials, consumer goods and foodstuffs. The development of regional electric power bases was the only addition to the 1939 list. Even so, the call for the development of regional machine-building and metallurgical industries voiced by Granovskii and Korobov was not admitted as a general principle of policy. To that extent the policy made some concession to the proponents of regional specialism and of the developed machine-building and metallurgical bases of the Old West.

The concern for complex development can also be seen in the details of the plan document. For example, thirteen of the sixteen union republics were to have developed their own cement industries by 1950. Of course, complex development was also to be applied to the Western and Central areas; for example the new areas of Moldavia were to be explored for coal and oil deposits and the iron ore workings in Tula and Lipetsk (Central economic region) were to be expanded. But, of the projects announced in the plan, proportionately more seemed to concern the eastern areas of the RSFSR and the Central Asian republics. To cite a few examples, Siberia was further to develop its own iron ore base to match its already vast coal output; the Urals (already a major producer of iron ore) was to develop its coal industry. Even the underdeveloped Far East economic region was to establish its own iron ore, coal, and therefore metallurgical, base. In Central Asia, the output of coal at Angrensk in Uzbekistan and of iron ore in Kazakhastan was to be rapidly expanded. Indeed coal output targets were set for all Central Asian republics, that for Uzbekistan being more than 300 times the 1940 output level! Coal, iron ore, and therefore metals, the bases of heavy industrial development, were to be produced in significant quantities in all the economic regions of the east of the USSR.

Another indicator of the commitment to complex development was the role assigned to locally controlled and co-operative industry. It was to produce 27.5 per cent of the USSR's industrial output in 1950. Much emphasis was laid on the need to develop local construction materials and fuel industries. This emphasis on local industry might also have reflected a concern for the administrative factor in regional policy. Co-ordination between enterprises in different industrial branches would surely be easier if both were controlled by the same republican Council of Ministers or local Soviet rather than each being run by separate ministeries in Moscow.

In the sphere of transport it was hoped to reduce the length of the average railway journey from 790 to 690 km over the Five-Year Plan period. Whilst industrial output in 1950 was to attain a level of 48 per cent above its 1940 level, freight turnover on the railways was to have risen only 28 per cent over the corresponding period. In other words, transportation of produce between economic regions was to be restricted. Within these regions, however, transport facilities were to be improved, especially in the east. Half the planned length of new railway track to be constructed over the period 1946–50 was to be laid in Siberia.

The ratio of railway transport turnover to industrial output in 1950 may have been planned at a lower level than in 1940, but the planned ratio was still slightly higher than in 1945. This reflects a pattern of regional development that permeated the whole of the Fourth Five-Year Plan. The proportions of 1940 (between both east and west and between transport turnover and output) were not to be restored, but then neither were those of 1945 (when the liberated areas were still producing at only one-third of their pre-war industrial output levels) to be retained. The east was to retain some of its wartime gains but not all. It was, however, due to undergo continued industrial expansion after the war; its industries were not to be run down to pay for the reconstruction of the western regions as some had seemed to suggest. In the words of one economist, writing in 1946:

> In the first post-war five year plan huge new capital construction will take place in Siberia, the Urals, and the Far East, where huge metallurgical and fuel bases and bases of the light and textile industries will be established. . . . As a result of these moves the role of the eastern regions in the economy of the nation will expand still further.[64]

The potential long-term benefits of investment in the underdeveloped areas of the east clearly influenced the compilers of the Fourth Five-Year Plan.

As can be seen from Table 3.2, some 14.2 per cent of planned capital investment in the USSR for the period 1946–50 was to be directed towards Siberia and the Far East and 5.7 per cent to the Central Asian republics.[65] One-fifth of capital investment was to be devoted to areas that had produced only 8 per cent of All-Union industrial output in 1940. Even in 1945, after the impact of the war, these areas had produced (respectively) only 11.6 and 4.3 per cent of total national output. On these indicators the position of the areas east of the Urals in the national economy was to be significantly improved not only in comparison with 1940, but even with 1945. This could also be seen in the targets set for the expansion of industrial output in all the Central Asian republics (see Table 3.2). All were to expand output over the period 1945–50 at a rate faster than or equal to the national average plan. Many specific projects for the development of new industries east of the Urals were listed in the plan. In addition the plan encouraged exploratory work to uncover more of the vast untapped mineral resources of the east.

TABLE 3.2. The Fourth Five-Year Plan: regional distribution of planned investment and output

	(1) Planned investment 1946/50	(2)	(3) Planned industrial output in 1950 (1945 = 100)	(4) Relative planned effectiveness[b] (ICORs) (USSR = 100)
	in billion roubles	as a percentage of USSR total		
USSR	250	100	161	100
Liberated areas	115	46	397	90.4
RSFSR[a]	146.6	58.6	147	90.2
Liberated areas	34.6	13.8	–	–
Siberia and Far East	35.6	14.2	–	–
UKRAINE	49.5	19.8	–	–
BELORUSSIA	6.95	2.8	580	95.7
MOLDAVIA	1.245	0.50	405	107.8
LITHUANIA	1.535	0.61	450	84.5
LATVIA	2.050	0.82	383	–
ESTONIA	3.500	1.40	411	–
GEORGIA	4.120	1.65	188	129.4
ARMENIA	1.420	0.57	226	87.7
AZERBAIDZHAN	5.900	2.36	151	159.8
UZBEKISTAN	3.900	1.56	177	67.6
KAZAKHSTAN	8.800	3.52	161	213.3
KIRGIZIA	1.200	0.48	172	150.3
TADZHIKISTAN	1.200	0.48	211	141.9
TURKMENISTAN	1.600	0.64	202	131.4

Sources: See Note 66 on p. 158.

[a] Including Karelo-Finnish Republic (incorporated into the RSFSR in 1956).
[b] ICOR = Incremental Capital/Output Ratio (see Appendix A).

However, nearly half of the planned All-Union investment was to be devoted to the continued reconstruction of the war-ravaged economy of the liberated areas in the west; and they had accounted for only one-third of Soviet industrial output in 1940 and only 10.8 per cent in 1945. Some new plants were to be built in these areas (especially in the Baltic states), but mostly the funds were to be expanded on reconstructing and re-equipping plants that had been partially destroyed during the war. Reconstruction was, however, not to be paid for by cutting back on new construction in the east. For example, the economies of the liberated areas of the RSFSR and of Siberia and the Far East were (according to the plan) to receive roughly similar allocations for investment (34.6 and 35.6 billion roubles respectively) over the five-year period; yet in 1940 these liberated areas had produced nearly three times as much industrial output as Siberia and the Far East. Even in 1945, after the ravages of war, these liberated regions were still producing nearly half the output of their unscathed eastern rivals.

The industry of the liberated areas of the USSR was to grow even faster than its eastern counterpart over the plan period. Yet, as one commentator noted in 1948,

> According to the Five Year Plan for 1946/50 the eastern regions will outstrip other areas of the country in terms of rates of growth. We will unhesitatingly continue our policy of redistributing industry towards the east, towards the national republics, and nearer to its source of raw materials and power.[67]

It was to be the unoccupied areas of the North and Centre and the westernmost of the eastern regions that were to make the sacrifices to pay for construction and reconstruction elsewhere. Only about 30 per cent of the USSR's investment funds were to be directed towards the unoccupied parts of the Central, North Caucasian, Northern, North-Western, Volga and Urals economic regions of the USSR, areas that had produced nearly two-thirds of All-Union industrial output in 1945 (and roughly 58 per cent even in 1940). These 'shortfall' areas can be divided into two groups—the old industrial areas of Centre, North-West and North Caucasus and the new areas of the Volga, Urals and the small Northern economic region; the latter group especially had benefited from wartime expansion (see Table 3.1). The precise balance of investment plans between these two types of region is, unfortunately, unknown. However, the detailed provisions of the plan did call for the development of a number of construction projects in the newly industrialised areas. For example, the Urals was to expand the coal-mining activities, develop a number of new chemical plants, see the 'forced' development of its hydro-electric power stations, and develop its own oilfields (especially in Bashkiria) and heavy machine-building capacity. The unoccupied areas of the Volga region were soon to become the nation's main source of oil and gas, and a massive hydro-electric scheme was to be begun on the Volga. In contrast the projects listed for the Central, North-Western and North Caucasian regions concentrated heavily on the reconstruction of their liberated zones rather than the construction of new plant in their unoccupied areas. This evidence is far from conclusive, but does suggest that it was the unoccupied parts of the large and well-developed Central and North-Western regions that were to forego some of the benefits of industrial growth to pay for expansion and reconstruction elsewhere.

Yet it was precisely in such industrially developed areas that economic policy-makers might have expected more rapid returns on investments, in the form of increased output. This argument applied *a fortiori* to many of the least damaged of the liberated areas (such as parts of the Baltic, which had been relatively well treated by the Nazis, and areas like the Moscow coal basin that had been occupied for only a short time). In such zones existing facilities and equipment could be renovated or adapted to productive use at a fraction of the cost of any new construction. This was indeed what the compilers and authorisers of the plan themselves expected. As can be seen from Column 4 of Table 3.2, the planned yield (in terms of increase in output over the years 1945–50) of each unit of investment was some 10 per cent higher in the liberated areas than in the USSR as a whole.[68] Much the same was true of the RSFSR, whose industry was still concentrated in liberated and unoccupied but old industrial areas, and of the other liberated Union republics, such as Lithuania and Belorussia. In contrast the yields expected from investment in most of the Central Asian republics (excepting only Uzbekistan, much the most industrially developed) and probably in the eastern regions of the RSFSR over the five-year period were significantly lower. In spite of this, substantial investment funds were to be directed to these areas. In other words, the pre-war policy of developing new *ochagi* in the east was to be continued; new industrial construction was to be undertaken in areas where it might yield less in economic terms over a five-year period than in other regions. The rationale for such a policy must have been in terms either of expectations of better longer-run economic returns, or of non-economic returns. Amongst these non-economic returns, considerations of nationalities policy must have weighed at a time of continued guerrilla warfare in the western Ukraine and after notably high abstention rates (about 10 per cent) in Lithuania and Estonia in the February 1946 elections to the Supreme Soviet.

The regional policy of the Fourth Five-Year Plan therefore had as its first priorities minimising transport costs and maximising long-run returns and non-economic returns on investments. Besides the demand for maximisation of short-run returns, the other criteria that received relatively little recognition in the plan were the dictates of defence and administrative efficiency. Obviously the USSR felt much more secure in its position in the world in 1946 than it had in 1939, this was evidenced by the low priority Voznesenskii accorded to defence requirements in his speech presenting the plan.[69] The question of administrative efficiency and especially of co-ordination of productive effort within each region was largely ignored. As has already been noted, the plan was based on the old economic regionalisation; worse still, no administrative units had been established in these economic regions to supervise their development or even to check on the implementation of this regional policy. That role was left by default to a motley collection of regional officials from Gosplan, and from party and Soviet apparati, who (as we shall see) found it increasingly difficult to enforce the policy against the wishes and interests of the central ministries.

The Soviet leadership in the months immediately after the announcement of the Five-Year Plan promulgated several decrees on various aspects of regional policy that again emphasised the need to promote industrial growth in the east. For example, decrees of August 1946 and May 1947 sought to encourage labour to move to the east.[70] The first decree provided for the payment of a 20 per cent bonus to all production workers in major branches of heavy industry in the Urals, Siberia and the Far East. As a further inducement, hundreds of thousands of new houses and flats were to be built in these areas by the end of 1947. The impact of this decree was to be fairly widespread: 824,000 workers were to receive the pay bonuses, probably between half and one-third of all industrial workers and employees in these areas. The second labour decree (which reorganised the whole labour recruitment system) promised large cash bonuses, special allocations of consumer goods, and travel expenses for workers willing to move from other areas to take up employment for one to two years in these eastern areas of the RSFSR. Clearly a conscious attempt was being made to ensure an adequate supply of labour to these areas.

Of the annual plans of this five-year period, only that for 1947 was published in any detailed form. As Table 3.3 shows, of the funds to be allocated for investment in the national economy, some 41 per cent was to go to the liberated areas and 5.2 per cent to Central Asia and Kazakhstan. The overall regional balance of investment was to be much as had been envisaged a year before in the Five-Year Plan.

The leadership's priorities in regional policy had changed little by March 1947, except perhaps in regard to administrative problems. Some recognition was given to the need to provide separate administrative units for different regions. First, during 1946 a few ministries (notably those running the coal, oil and fisheries industries) were split into two, each new ministry controlling one regional 'half' of the industry concerned. In time this might have amounted to something rather more than 'renaming glavki',[72] but the ministries were reunited over the following two years as the reconstruction of western industry was substantially completed.

TABLE 3.3. The annual economic plan for 1947: regional aspects

Area	*Planned investment in 1947 (R million)*	*% of USSR*	*Planned industrial output in 1947 (1946 = 100)*
USSR	50000	100	
Liberated areas	20500	41	137
RSFSR[a]	26500	53	107
UKRAINE	9770	19.5	120
BELORUSSIA	1045	2.1	136
MOLDAVIA	173	0.35	116
LITHUANIA	202	0.40	140
LATVIA	277	0.55	119
ESTONIA	570	1.1	121
GEORGIA	890	1.8	110
ARMENIA	350	0.70	117
AZERBAIDZHAN	1000	2.0	106
UZBEKISTAN	800	1.6	122
KAZAKHSTAN	1200	2.4	106
KIRGIZIA	180	0.36	113
TADZHIKISTAN	130	0.26	117
TURKMENISTAN	280	0.56	110
KARELO/FINLAND	270	0.54	162
UNDISTRIBUTED	6363	12.7	

Source: See Note 71 on p. 158.
[a] Excluding Karelo-Finnish SSR.

Potentially more useful was the reorganisation of Gosplan's regional department in January 1947. In the context of a general attempt to improve the planning or proportions in the economy, a sector for each Union republic was established within this Department of Regional Distribution. The aim was to improve co-ordination within each republic and thereby boost the development of each republic's economy and ensure the complex development of each economic region.[73] However, each regional section had had only thirty to forty staff in 1940, not enough to really carry out its tasks. In any case, it was the production-branch-divided central ministries that were actually charged with administering industrial enterprises, not the departments of Gosplan. Such piecemeal attempts at administrative reform were more an indication of the leadership's wishes than a practical way of ensuring the implementation of the regional policy.

Regional Policy Discussions 1946–52

The fact that the leadership arrived at and announced various decisions on regional policy during 1946 and 1947 did not preclude further debate on the issue. No one would, of course, openly challenge the advisibility of the plan directives or decrees, but commentators could seek to influence the future trend of regional policy by discussing the non-fulfilment of the plan directives; if the leadership's decrees were not being fulfilled, the reasons could be sought in private and public debate. Participants in these discussions would also claim to be discussing new problems and features of particular areas; if, for example, a natural disaster (such as the drought of 1946) occurred, or a large new mineral deposit was found, was it not legitimate to reopen discussion of regional policy in the light of these new developments? In addition, debate on some aspects of regional policy was permitted on the grounds that certain issues were quite legitimate spheres for academic research and debate; in particular, debates over a new regionalisation and over the criteria to be used in evaluating new investment projects affected the trend of regional policy during Stalin's last decade. These, then, were the three main means by which those seeking to affect regional policy formation could circumvent one of the main requirements of the Soviet doctrine of democratic centralism, that discussion should cease on the issue after a decision has been made.

The main practical impact of their discussions in the sphere of regional policy was to re-emphasise the right of the political leadership to take decisions favouring the east that could not be justified on short-run maximisation grounds. At the same time the leadership did (implicitly) admit the right of those economists and others who were advocating some calculation of the returns to be expected from investments to influence regional policy decisions.

The first debate involved not only academic economists, but also more journalistic commentators and interested regional officials, all of whom sought to press the claims of particular areas on policy-makers. In the mass of commentaries on the new Five-Year Plan that appeared in publications in 1946, several authors stressed its provisions on the eastern regions, probably with as much an eye on influencing the implementation of plan policy as on affecting future regional policy decisions. For example, Feigin writing in *Bol'shevik*, stressed the long-run specialisation, transport economy and defence arguments for further developments of the Near East. He argued that the entire capital cost of further development of the iron and steel industry in the Kuzbass and Magnitogorsk could be recouped within ten to twelve years solely by the savings in transportation inputs to the national economy.[74] Arakelyan commented that, as a result of new capital construction in the eastern regions of the RSFSR under the new plan, their '. . . role in the economy of the nation will still further expand'.[75]

The western areas of the USSR were also not without their advocates. Volodarskii emphasised the need to concentrate resources on the liberated areas, as it was the most badly damaged enterprises that had not been reconstructed during the war that should now be placed high on the list of priorities. The reconstruction effort would therefore need more resources than some might think, a point echoed by another author with respect to the need for housing and other services in liberated areas of the RSFSR.[76]

Even after the post-plan publicity had died down articulations of regional interests continued in the press and at various conferences. The eastern regions' claims on national resources were most notably advanced at a series of conferences in 1947 and 1948 on the future development of the productive forces of various eastern areas. These were organised under the auspices of Gosplan by the Institute of Economics of the Academy of Sciences in co-operation with local party and Soviet organs. Their findings were to form the basis of a new Twenty-Year Plan. Those participating included academic economists, planners and local party and Soviet officials—both checkers and line officials—of the command economy. For example, the September 1947 Conference on Tadzhikistan had an organising committee consisting of two nationally known academics (Bardin and Shevyakov), a local party secretary (Gafurov) and the chairman of the republic's Council of Ministers (Rasulov). Such conferences not only brought together members of different organisations, but also provided forums for the articulation of their common interests and opinions in the national press. In previewing the conference on the future development of Irkutsk oblast' in August 1947, the chairman of the local Soviet committee stressed the richness of the raw material reserves in the area.[77] The Vice-President of the USSR Academy of Sciences, I. P. Bardin (who was partly responsible for the convening of this and similar conferences), called for 'the transformation of a far-distant, sparsely-populated, still insufficiently . . . developed region into one of the most important industrial centres'.[78]

The economist Shevyakov also called for the complex development of East Siberia.[79] However, the projects listed by the conference as primary tasks were not realised for many years; the local Cheremkhovsk coal basin, for example, was not opened up until the 1960s. Clearly such conferences had only a limited impact on immediate economic realities.

In April 1948 a similar conference was convened in Molotov oblast. In his opening speech P. A. Khromov pointed to the rise in labour productivity that could be secured by concentrating industrial development nearer to raw material sources in the east.[80] In view of such comments by members of the USSR Academy of Sciences, it is hardly surprising that its Institute of Economics re-established its sector for economic regions, under the auspices of Gosplan. Its tasks were to work out prospective plans for the regions, to promote new towns and industrial centres, and actively to participate in the conferences described above.[81] In other words, it was to act as a sort of pressure group for eastern expansion, providing information on those regions (gleaned at conferences) to Gosplan and the Council of Ministers; that is, to the staff officials of the command economy.

Conferences on developing the potential of many western areas were hardly necessary. However, local officials in particular were not slow to propose the further expansion of their areas. For example, a local trades union official called for the rapid expansion of Leningrad's own electric-power base in 1946.[82] Molotov, speaking to a session of the Ukrainian Supreme Soviet in 1948 called for further help to the new western oblasti of that republic from the rest of the USSR.[83] The chief engineer of a Donbass coal combine claimed in 1948 that the possibilities for the future development of the Donbass coal mines were 'unlimited. Of this there is no doubt'.[84] The Donbass had less than 3 per cent of the USSR's coal reserves!

Supreme Soviet sessions and Party Congresses also acted partly as forums for regional policy discussions. For example, the Finance Minister, Zverev, considered a number of requests for over-plan appropriations for particular areas in a Supreme Soviet budget debate in March 1949. Some, such as a new railway in the Fergan Valley in Uzbekistan he reviewed favourably; this project did indeed find its way into the Fifth Five-Year Plan directives in 1952. Others, such as the request from a Belorussian deputy for aid in restoring local mills, Zverev rejected outright.[85] Such articulation of regional interests by local officials was an accepted part of the policy-formation process. The speeches of delegates from the regions to the 19th Party Congress were also marked by their concern with local matters and the claims made on higher authorities. N. S. Patolichev, First Secretary of the Belorussian Party and soon to join the All-Union Praesidium, demanded that 'these favourable opportunities for further development of Belorussian industry must be more fully taken into account both by the republic's administrative organisations and by planning bodies and All-Union ministries.'[86] Such points were, however, made in an attempt to influence decisions on specific regions rather than regional policy in general.

The debates amongst economists and economic geographers were, in contrast, very much concerned with issues of general policy. Indeed academics were specifically encouraged to take an active part in policy discussions at this time. An editorial in *Bol'shevik* in 1946 pointed out that 'the scientific worker in our country is a social activist. He cannot be apolitical', and accused Soviet economists of concentrating on historical rather than current problems and on churning out facts and figures rather than analysing.[87] These criticisms were repeated by I. A. Gladkov in a 1947 review of the literature on the new Five-Year Plan, and by Ostrovityanov in at least two swingeing attacks on his fellow members of the newly reorganised Institute of Economics in 1948.[88]

From one point of view, of course, these criticisms, in the general spirit of the 'Zhdanovshchina', were designed to suppress discussion of leadership policies rather than encourage it. Ostrovityanov, for example, had been appointed head of the Institute of Economics to remove the influence of such dissident economists as Yevgeny Varga. However, his criticisms of other economists for lack of 'bravery' in confronting current issues were rather more reasonable than the attacks on Varga and on his ideas on the possibilities of non-revolutionary progress to socialism in the West. Many of the other works published at this time were overly historical or factual. This cannot be attributed solely to economists' fears of the political consequences of analysing current problems, for those consequences, as in the cultural sphere, were not too severe at this time. Varga, for example, defended his position rather than engaging in the fundamental self-criticism called for by his opponents. In spite of this, his institute survived, although now subordinated to the Institute of Economics. Varga himself was demoted but not dismissed, let alone arrested. Several speakers at these sessions accused Ostrovityanov in particular of being purely negative; A. I. Kats, for example, argued that scientific research could not achieve its aims if directed by the sort of 'general formulas' Ostrovityanov was proposing. Kats himself, having had an article rejected by the institute's journal, got it published by the party journal *Bol'shevik*![89]

If academics still preserved some freedom of discussion in the delicate field of the economics of capitalist countries, they enjoyed far more liberty in discussing domestic problems like regional policy. One member of the new institute, D. I. Chernomordik, attacked his fellow economists for failing to study the problems of complex development.[90] But the truth was that economic geographers in particular were still bogged down in discussions of which regions they should use as the basis for such research. The regionalisation debate that had produced the clash in Gosplan in 1945 was by no means dead.

On one side of the argument were ranged those who sought to use existing administrative boundaries as the basis for detailed study of the regions. Their opponents advanced the view that the regions should be delineated according to natural conditions and/or existing production links between enterprises. The policy implications of this latter view were that plans for complex development should be drawn up according to the existing distribution of industry, communications and raw materials. That would lay stress on the need to maximise shorter-run returns in taking location decisions and favour the more developed regions.

G. S. Nevel'shtein attacked Gosplan's current raionirovanie on the grounds that its regions lacked unity in that they were based on administrative boundaries rather than production links. N. N. Kolosovskii pointed to the resultant poor co-ordination of economic activity within these regions. A compromise solution was proposed by M. Vol'f. He suggested that existing administrative boundaries could be used for nationalities policy, current (short-run) planning and the publication of regional statistics, whilst boundaries based on production links and natural resources could be used as a basis for teaching, research, and (by implication) long-run perspective planning.[91]

In reviewing the work of these authors Yu Saushkin accused Vol'f of divorcing teaching and study from practical political problems.[92] The ideas of Nevel'shtein and Kolosovskii he attacked as too static. Boundaries of regions should, according to Saushkin, be set in accord with plans for the future development of productive forces. Raionirovanie was not the technical engineering task which Nevel'shtein described, but one involving essentially political considerations, notably nationalities policy. In other words, Saushkin emphasised the importance of long-run specialisation and nationalities factors to regional policy, arguments that tended to favour the eastern regions.

At the October session of the Institute of Economics V. F. Vasyutin elaborated similar arguments. He attacked the likes of Kolosovskii, Nevel'shtein and Vol'f for ignoring the fact that 'regional distribution is not independent of production relations'.[93] In an economy organised along socialist lines location decisions were a matter of politics, not of minimising short-run costs. In an attack on the agricultural economists Nemchinov and Kolesnev (and, incidentally, on Saushkin) he accused them of being subject to Webereian influences in their stress on natural factors in determining location decisions. It was for this error that N. N. Baranskii was dismissed from the editorial seat of the journal *Geografiya v shkole*. Even so, some were concerned to keep the debate above the level of factional controversy. Saushkin, for example (whose comments above were admittedly rather self-critical) termed one of Nevel'shtein's articles 'useful' and called for it to be circulated more widely (the first printing had reached only 400 copies).[94]

The public debate over regionalisation did, however, begin to ease off after 1948 after a resolution of the Institute noted that 'On the economic geography front the influence of idealism and of vulgar political economy has still not been eradicated'.[95]

The debate over criteria for investment decisions that had begun before the war (when Novozhilov first published his views) was still going strong in 1950. The debate centred on the extent to which prospective yields and operating costs of investment projects should be discounted against their initial (capital) costs when choices between alternative projects had to be made.[96] The practice in many industries was simply to compute a rough expected 'coefficient of effectiveness' of capital. For example, in the railway sector projects were compared in terms of initial cost and of their operating costs and yields over a ten-year 'period of recoupment', a 'coefficient of effectiveness' of 10 per cent. V. V. Novozhilov and A. Lure attacked this practice on the grounds that it involved crude averaging processes.[97] They advocated the use of Western marginal concepts; the marginal efficiency of capital was the relevant criterion for choosing between projects.

Such 'bourgeois' influences were attacked by economists such as T. S. Khachaturov, P. Mstislavskii and S. G. Strumilin[98] on the grounds that Novozhilov and Lure were seeking the reintroduction of capitalist rate-of-interest criteria. These three economists all emphasised the problems of applying such monetary criteria in a system where prices were administratively fixed (and did not therefore necessarily reflect costs). That criticism applied just as much to existing practices as to the marginal theories of Novozhilov.

Perhaps the most ingenious solution was that of Strumilin. He argued that both initial and operating costs should be calculated in terms of embodied labour. Even if this were done, however, the value of capital (in terms of embodied labour) would still fall over time, as the cost of replacing it would fall as the productivity of labour rose. In other words, the net yield of any capital expenditure would fall from year to year and so could be said to reach a finite sum. The total cost of each project (combining initial and operating costs) could therefore be assessed in terms of embodied labour.

Mstislavskii attacked even this seemingly unimpeachable Marxism as a disguised form of capitalist interest-rate criterion. He argued that the choice between investment projects, especially between different sectors and regions was essentially a 'politico-economic' matter, not the technological exercise envisaged by Strumilin. There was no reason to suppose that the value of capital would decline over time at the same rate in different regions and sectors. Such an assumption ignored the social costs and benefits ('concomitant costs') of projects. In summary, Mstislavskii was advocating that investment choice (and thus location decisions) should be left to planners and, ultimately to politicians, not economists. That position was echoed amongst economic geographers by Vasyutin and Karnaukhova, who both attacked Nemchinov and Kolesnev for using any kind of 'cost-minimisation' criteria for choosing between projects.[99] As Vasyutin pointed out, the use of any such criterion would discriminate against the initially costly projects in the less developed areas of the east.

Both Mstislavskii and Vasyutin invoked the name of Stalin in their attacks on their fellow economists and their contributions were widely regarded in the West at this time as key parts of the 'Zhdanovshchina'.[100] Yet their attacks were not simply personal condemnations, and their main opponents continued to write and to stand up for their points of view. Indeed, the final word seems to have belonged not to Mstislavskii in 1948–9 but to Khachaturov, addressing a meeting of economists in 1950.[101]

Khachaturov had, four years earlier, come out in favour of a criterion involving calculating the average cost of a project over a limited period of time,[102] but now accepted both Strumilin's and Mstislavskii's criticisms. Yet, argued Khachaturov, some calculation and comparison of initial operating costs had to be made. The economist must play a part in choosing between projects (and thus in making location decisions). The social costs mentioned by Mstislavskii were important factors to be taken into account by planners, but, equally, so were the cost-effectiveness calculations of economists. Economists, therefore, could and should influence investment decisions, but the final choice should lie with those who took the final decisions on the plan.

In restoring the economists' claim to be involved in the process of policy-formation (including that of regional policy) Khachaturov seemed to represent something of a consensus amongst his colleagues. The exact method of combining initial and net operating costs and, in particular, the extent to which the latter should be discounted over time was, however, still open to dispute. As far as regional policy was concerned, the bias for or against the eastern areas depended very much on the rate of discount or period of recoupment to be used; the lower the rate (or the longer the time-period), the more the short-run benefits of investment in the more developed areas of the Old West would be offset by the long-run returns from projects in the more backward east. In any case the paucity of reliable statistics and the terrible state of Soviet projecting work at the time made the choice between rather academic. Even so the drive for quicker returns on investments (part of the attack on 'gigantomania' mounted in 1949–50) can have done little to persuade planners and administrators to locate new projects in the east.[103]

These discussions amongst economists and various interested officials of the command economy were part of a continuous process. Just as advisers and administrators well outside the leadership circle of the Politburo and Secretariat had previously sought to influence the content of the Fourth Five-Year Plan, so they now concentrated their energies on influencing future decisions, and especially the new Five-Year Plan due to go into effect from January 1951. This plan did not appear until October 1952 and was distinguished by its lack of commitment to the regional policy of 1939 and 1946. This may have been partly due to the impact of these discussions, but was perhaps more attributable to the problems experienced in implementing the old policy. In the longer run, however, the advocates of eastern-based and complex development had far more success in influencing the decisions of the post-Stalin leadership.

Before analysing the decisions of 1952 and after it would be useful to note the role played by academics in the discussions. Academic debate was far less restrained and far more public than differences of opinion amongst line officials and political leaders. The two sets of discussions were, however, directly linked. Academics were expected to provide expert advice on matters like regional policy to staff officials and the political leadership. It was this role that Khachaturov was restating in 1950. It was also embodied in more formal connections such as the subordination of the Institute of Economics to Gosplan in 1948. The debates cited above show that this move, if intended to stifle discussion, was not very successful. It can also be interpreted as a move to give decision-makers and planners better access to expert advice and to halt the drift towards irrelevant theoretical work in economics. The convening of conferences was a less consistent but none the less effective way of linking academics, officials and leaders. In the sphere of regional policy at least, academics and journalistic commentators acted both as publicists for debates and as staff advisers in the command economy.

Regional Policy Decisions 1952 and After

The Fifth Five-Year Plan was the swansong of the Stalin era. The General Secretary died six months after its announcement and policy changes followed in many spheres. The plan document was noticeably less informative than its predecessor of 1946, partly due to the enactment of decrees making the publication of much economic data illegal.[104] As approved by the 19th Party Congress the new plan made none of the recommendations for individual republics that had been published in 1946. This may well have been due to the requirements of secrecy. Gosplan continued to break down overall annual targets into regional divisions on the basis of the criteria of the Third Five-Year Plan.[105] The new plan reiterated the need to 'improve the geographic distribution of new industrial construction' and stressed the priority of locating plants near to sources of fuel and raw materials and of cutting transport costs.[106] Railway transport turnover was to increase by only 35 to 40 per cent whilst industrial output was to grow by 60 per cent over the five-year period. In addition a number of large construction projects in the east were announced (such as hydro-electric power stations on the Kama and Irtysh rivers in Siberia). After the Congress debate on the draft plan the commitment to expand iron and steel production 'especially in the eastern regions' was extended to the railway and coal sectors.

The new plan merely reiterated existing policy; it did little to reinforce it. The major problem, as we shall see, was that it was not being carried out as intended. The advocates of the east and of complex development criteria had a definite impact on policy formation but not on its implementation. Several delegates to the Party Congress itself complained that the ministries were not implementing the regional policy of 1939, 1946 and 1952. For example, the First Secretary of Khabarovsk Kraikom (in the Far East) complained to the Congress that the Ministry of the Forestry Industry had not remedied defects in its location policy and asked the party's central committee and the government to put pressure on the ministry.[107] The Kirgizian party secretary accused the central ministries of coal and railways of failing to develop a local coal basin as prescribed in the Fourth Five-Year Plan.[108]

The discussions of the 1940s were, however, to bear more fruit after Stalin's death when the problems of implementing regional policy were officially recognised. In 1954 the arguments that had been going on in Gosplan since 1945 acquired a new impetus.[109] This led to the *Sovnarkhoz* reforms of 1957 which placed some 70 per cent of Soviet industrial output under regional economic councils (Sovnarkhozy). For the first time regional policy was to be carried out by regional administrative agencies. Those production-branch ministries that survived the reform lost much of their day-to-day control over the economy and with it much of their power to frustrate the regional policy of the late Stalin era. These councils in turn showed themselves capable of frustrating leadership policy, albeit in a 'localist' rather than 'departmentalist' direction. The result was the re-emergence of the old branch ministerial system in 1965. The role of officials and advisers in the command economy in influencing regional policy has long outlasted the death of Stalin. The mistake often made by Western analysts is to assume that such influences only appeared after Stalin's death when the command economy model faced a more reformist political leadership.

In the Stalin era, as later, however, it was one thing to get a particular policy enshrined as a leadership priority in plans and decrees; it was quite another to ensure that the apparatus of the command economy obeyed its masters, even in such priority matters.

4 The Implementation of Regional Policy after the War

The Fulfilment of the Fourth Five-Year Plan

The output of Soviet industry as a whole grew rapidly over the Fourth Five-Year Plan period. It rose by some 88 per cent, 17 per cent more than had been envisaged in the plan document. The regional policy of the plan was, however, by no means fully implemented. As we have seen, the plan policy was one of complex development of all regions favouring the underdeveloped areas in the east and the new areas in the west of the USSR. Such a policy should have led to greater economy in the use of transport facilities, especially of the railway system. Yet neither the target for restraining total railway freight turnover nor that for cutting the average length of freight haul were attained in 1950. The former was exceeded by 13 per cent, a fact that might be accounted for by the 17 per cent overfulfilment of the industrial output plan. The average freight haul was cut, but only to 722 km., as against the planned 690 and the 1940 level of 700 km.[1] The demand for interregional railway transport continued to rise and was met, in great part, by raising operating norms rather than by investing in new track or equipment. It need only be noted here that in this crucial respect the regional policy of 1946 was far from perfectly executed.

Neither were the plans for the further development of the east of the USSR fully implemented, as Table 4.1 shows.

The industrial output of the Siberian and Far Eastern economic regions grew slower than the national average. That could be expected in view of the need to reconstruct industry in the West; what was surely not envisaged in 1946 was that the annual output of the Urals region would *fall* over the plan period and that that of the West Siberian region would grow only by 3 per cent per annum. The total output of the Urals, Siberia and the Far East rose by only 7 per cent over the period of the plan. As a result these

TABLE 4.1. Fulfilment of the Fourth Five-Year Plan's regional output targets

Region	*Actual growth of industrial output (1950 at 1945 = 100)*	*Percentage fulfilment of plan*
USSR	188	117
Liberated areas	390	98
RSFSR[a]	165	112
North	203	–
North-West[a]	388	–
Centre	189	–
North Caucasus	283	–
Volga	145	–
Urals	96	–
West Siberia	116	–
East Siberia	155	–
Far East	146	–
UKRAINE	442	–
BELORUSSIA	575	99.1
MOLDAVIA	468	116
LITHUANIA	478	106
LATVIA	645	168
ESTONIA	468	114
GEORGIA	195	104
ARMENIA	268	119
AZERBAIDZHAN	178	118
UZBEKISTAN	171	97
KAZAKHSTAN	169	105
KIRGIZIA	176	102
TADZHIKISTAN	204	97
TURKMENISTAN	164	81

Sources: See Note 2 on p. 159.

[a] Including Karelo-Finnish SSR.

areas produced only 19.1 per cent of Soviet industrial output in 1950 compared with 33.5 per cent in 1945 and 12.2 per cent in 1940.

The non-Russian areas of the east—Central Asia and Kazakhstan—fared rather better; the industrial output of these republics increased by 74 per cent over the plan period. However, this was still below the national average growth rate. In addition the output plans for three of the five republics were not fulfilled and those for Kazakhstan and Kirgizia only marginally overfulfilled.

The importance of the eastern regions to the national economy therefore declined over the plan period, a decline that was more marked than the policy-makers had proposed in 1946. Yet the targets for the reconstruction of western industry were also not met. The output of the liberated areas nearly quadrupled over the years 1945–50, but their share of All-Union industrial output in 1950 was still significantly less than planned; 1940 output levels were not greatly exceeded.

It was in the remaining unoccupied areas of the RSFSR and in the Baltic and Caucasian republics that output targets for 1950 were comfortably surpassed. The output of all the new areas[3] that had been added to the west of the USSR in 1939–40 expanded by more than five times over the plan period. Some concern for nationalities policy could perhaps be imputed to those responsible.

The old industrial areas of the RSFSR continued to grow rapidly. Yet large parts of these regions had never been occupied by the Nazis, were not populated by non-Russian races, and were scarcely in need of many new branches of industry to ensure their complex development. Neither were there great new sources of raw materials waiting to be exploited in the west of the RSFSR (the Mosbass being a notable exception). The only possible rationale for the continued expansion of these areas was in terms of short-run specialisation and maximising short-run returns on investment.

The industrial output of the unoccupied zones of the RSFSR west of the Urals and the Caucasian republics more than doubled over the plan period, having suffered very little from the ravages of war. For example, the heavily industrialised Kirov oblast' not only witnessed an industrial growth rate of 89 per cent over the war years, but also a rate of 65 per cent over the next five years! The policy of 1945 had been to let these developed industrial areas run down and to expand the new ones in the east. In the implementation of the policy, however, such long-run criteria took second place. The focal point of Soviet industrial development remained west of the Urals.

Regional Policy in 1951 and 1952

The lack of specific target figures for the various republics and regions of the USSR make it very difficult to assess the level of fulfilment of regional policy in the Fifth Five-Year Plan period. In many ways such an exercise is pointless because Stalin died only six months after the plan was approved.

TABLE 4.2. Industrial growth rates by regions and republic 1951–3

	Output in 1951 at 1950 = 100	*Output in 1952 at 1951 = 100*	*Output in 1952 at 1950 = 100*	*Output in 1953 at 1950 = 100*
USSR				146
RSFSR				142
of which:				
Centre	116	110	128	145
North	111	105	117	123
North-West	118	113	133	149
North Caucasus	116	110	128	145
Volga	116	110	128	141
Urals	114	112	128	140
West Siberia	116	112	130	143
East Siberia	117	111	130	139
Far East	117	108	127	135
UKRAINE				150
LITHUANIA				182
LATVIA				150
ESTONIA				154
GEORGIA				138
ARMENIA				148
AZERBAIDZHAN				128
UZBEKISTAN				137
KIRGIZIA				144
TADZHIKISTAN				147
TURKMENISTAN				145
KAZAKHSTAN				140
BELORUSSIA				156
MOLDAVIA				183

Sources: See Note 4 on p. 159.

However, over the period between the end of the previous plan and the approval of its successor little progress was made in implementing regional policy. As Table 4.2 shows, industrial growth rates amongst the various regions were remarkably uniform in 1951 and 1952. There was virtually no redistribution of industrial growth within the RSFSR over this period. If anything, it was the more backward North and Far Eastern regions that were the slowest growing and the highly developed North-West the fastest! Over the period 1950–3 the lagging behind of the North and the easternmost economic regions of the RSFSR was even more pronounced. Of the union republics the largest in Central Asia—Uzbekistan and

Kazakhstan—grew slower than the national average. It was the liberated republics of the Ukraine and Belorussia and the new areas of the Baltic and Moldavia that experienced the greatest industrial growth in the first years of the Fifth Five-Year Plan. In Stalin's last years there was no redistribution of Soviet industry towards the underdeveloped east. The only major change after 1950 was in the relatively rapid rise of liberated but otherwise developed western areas such as the Ukraine.

The only indicator of complex development specified in the published plan in October 1952 was the requirement that railway transport turnover should increase by only 35–40 per cent over the period 1950–5 whereas industrial output was to rise by 70 per cent. In other words the leadership wanted to cut the transportation/output ratio by about one-fifth in five years. In 1953 the railway network carried 32.5 per cent more traffic than it had done in 1950; in the meantime industrial output had risen by 46 per cent. In other words in three years the ratio had been reduced by more than 9 per cent. Even so it seems unlikely that the plan target would have been met by the Stalinist Command Economy.

In 1951–3, therefore, as in the previous five years, there was little or no redistribution of industrial growth in favour of the east. The policies of complex development that dictated heavy investment in these more backward and slower-yielding areas were not implemented.

Sources of Overfulfilment and Underfulfilment

Why were output plans for some areas significantly overfulfilled and those for others barely or even under-fulfilled? If the production functions envisaged by the planners in 1946 bore any relation to regional realities, then there must have been an oversupply of some factors of production to the more successful areas and a shortfall in supplies to the less successful areas. If we can identify these surpluses and shortfalls in particular factor markets, then we can begin to allot responsibility for the distortions we have noted in the implementation of regional policy. If the west received better supplies of labour, capital, fuel or raw materials than the east or simply had better managers for its plants, then we can place the blame for the failure to carry out regional policy on the sectors of the administrative system that controlled these supplies. As we shall see, suffering managers and officials themselves were often not slow to point out the guilty parties.

We will now analyse the supply situation in four major markets—capital, labour, material inputs and transport—and the administrative problems faced by the regions.

Capital Allocations and their Utilisation

In the case of the capital market, we must assess the extent to which capital investment funds were distributed amongst the regions in the proportions laid out in the plan. In particular, were centrally allocated investment funds diverted from the eastern regions by those charged with the execution of regional policy?

Table 4.3 seems to suggest that, on the contrary, it was the liberated areas of the Ukraine, Belorussia and the Baltic that received rather less

TABLE 4.3. Actual capital investment and its effectiveness by regions of the USSR during the Fourth Five-Year Plan period

Region	*Share of USSR Total Investment in national economy per cent*		*Share of USSR industrial investment per cent*	*Relative effectiveness of capital investment (ICOR) (USSR = 100)*	
	Planned	*Actual*		*Planned*	*Actual*
USSR	100	100	100	100	100
Liberated areas	46	n.a.	n.a	90.4	–
RSFSR[a]	58.6	62.7	61.9	90.2	101.1
Liberated areas	13.8	n.a.	n.a.	–	–
Siberia and Far East	14.2	13.4[b]	n.a.	–	–
UKRAINE	19.8	19.2	21.7	–	102.7
BELORUSSIA	2.8	2.3	1.6	95.7	114.4
MOLDAVIA	0.5	0.5	0.2	107.8	135.8
LITHUANIA	0.61	0.5	0.3	84.5	92.3
LATVIA	0.82	0.9	0.5	–	–
ESTONIA	1.4	0.9	0.7	–	–
GEORGIA	1.65	2.0	2.0	129.4	209.1
ARMENIA	0.57	0.7	0.7	87.7	117.2
AZERBAIDZHAN	2.36	2.5	3.2	159.8	160.5
UZBEKISTAN	1.56	1.9	1.7	67.6	128.4
KAZAKHSTAN	3.52	3.7	3.3	213.3	281.7
KIRGIZIA	0.48	0.5	0.4	150.3	214.3
TADZHIKISTAN	0.48	0.5	0.3	141.9	227.7
TURKMENISTAN	0.64	0.8	0.8	131.4	377.5
UNDISTRIBUTED	4.2	0.4	0.7	–	–

Sources: See Note 6 on p. 159.
[a] including Karelo-Finnish SSR.
[b] excluding Kurgan oblast'.

than their planned share of USSR capital investment funds, whilst Central Asia and Kazakhstan benefited rather than suffered in the implementation of this part of the plan.

The major beneficiary was the Russian republic. However, the share of investment funds that reached the three easternmost economic regions (Siberia and the Far East) was probably slightly *less* than planned. Capital allocations to the other major eastern economic region, the Urals, may have exceeded the (unknown) planned level, but the available evidence suggests that this is unlikely. The Urals received only 10.4 per cent of USSR capital investment funds during the Fourth Five-Year Plan, compared to 16 per cent during the war and 8.5 per cent even during the (uncompleted) Third Five-Year Plan period. Indeed, all the eastern regions together accounted for a proportion of Soviet investment funds over the period 1946–50 (30.8 per cent) that was actually less than in the last pre-war plan period and only a little higher than in the Second Five-Year Plan period (29.2 per cent); that is, long before any overall regional policy had been approved.[5] However, whether this apparent lack of progress in the development of the east was due solely to biases in plan implementation rather than plan formation must remain open to doubt.

In summary, therefore, the pattern of allocation of investment funds over the post-war period showed no great deviations from the planned proportions. Such deviations as there were in fact beneficial to some of these eastern areas. Those responsible for checking on the fulfilment of the plan in this respect (notably Gosplan's capital construction department and republican governments who controlled investment in local industries) cannot therefore be held to account for the failure to implement many of the main features of the 1946 regional policy.

The allocation of investment funds (*kapital'nye vlozheniya*) is, however, quite another matter from the use to which those funds are put; that is, from the actual effectiveness of those funds in producing usable productive capacity. Millions of roubles invested in unfinished plant or in plant that lacks essential current inputs (labour, raw materials, etc.) are of little immediate benefit to the economy of the region concerned. Such supply and construction problems were rife in the Soviet economy in the immediate post-war years.

It has been envisaged by the planners that capital investments should show a higher yield in the more developed western and central areas than elsewhere. But, as Table 4.3 shows, the variations in actual yield were even greater than planned. Capital investments yielded the highest returns in the Baltic States and the liberated republics of the Ukraine and Belorussia and the lowest (over the plan period) in Central Asia and Kazakhstan.

Perhaps more significantly, capital investments were used more than twice as unproductively than in the USSR as a whole in four of the five eastern republics compared to a *planned* ratio of more like 3:2. In the fulfilment of the plan, capital allocations were used less efficiently than the planners had hoped in most areas, but far more so in the eastern than in the western non-Russian republics.

According to this aggregate data the Russian republic itself was one of the most productive users of capital investment in both planned and actual terms. Yet the actual yield of funds invested in the Urals and Siberia was little more than one-tenth of the USSR or the RSFSR average yield. Even if the data for the Urals is excluded, yields were at best less than one-third of that average. Although there are no data available as to the planned level of capital effectiveness in these areas, one must conclude that investment allocations were used far less effectively in the Urals, Siberia and the Far East than in the older industrial regions of the RSFSR. It is extremely unlikely that variations in yields between regions of up to twenty times were actually *planned* for. As the Siberian economic historian G. A. Dokuchaev has noted:

> The Soviet government expended a not inconsiderable sum [on capital investment in Siberia and the Far East], skilful economical and rational use of which could have produced excellent results. However, these were not achieved everywhere. The basic shortcoming in construction was the low quality of the work and the extraordinary prolongation of the periods of completing construction projects. Allocated funds, especially in the first years of the five year plan, were not fully utilised, often only to a level of 50 to 60 per cent.[7]

To summarise, the relatively poor performance of the eastern areas during the Five-Year Plan was not so much due to any shortage of capital allocations as to the fact that those allocations were not used to their full effect. This was in turn due to shortages of such inputs as qualified workers and construction materials, as well as poor planning. Indeed, one could broaden Dokuchaev's analysis to argue that it was such shortages of supplies of human and material resources (to both construction sites and already operating plant) that were decisive in frustrating the growth of eastern industry after the war, thereby preventing the enactment of the leadership's regional policies. This conclusion is based not only on data concerning distribution of these inputs to various regions, but also on the mass of complaints about poor supplies voiced in the press and at conferences and meetings by the officials concerned. It is these complainants who so often lay the blame for the supply situation at the doors of the All-Union branch-ministries. It has already been argued that these ministries had ample powers to determine the regional distribution of supplies. It will be argued below that they had little interest in ensuring that supplies went eastwards rather than westwards or stayed in the old industrial regions around Moscow and Leningrad.

If the aim of each ministry was to maximise the output of its particular branch of industry, would not that aim be best served by directing as many funds and supplies as possible to the areas where they might yield the most rapid increases in physical output? In the context of the Fourth Five-Year Plan period that clearly meant diverting resources to the faster-yielding construction projects and industrial enterprises of the old industrial regions rather than to the infant industries of the east (or even to the worst-damaged areas of the west). Faced with conflicting demands both to maximise branch output and to redistribute productive forces over the plan period, the ministries had to choose between meeting one requirement or another. They generally chose to maximise the output of the branch to fulfil *their* own plans rather than of particular regions, to help fulfil the plan for which some other organisation (a party organ or a Gosplan department) was responsible. Even if specific priorities of the leadership (such as the western areas or the Kuznets Basin in 1947–8) were announced, the ministries would still keep the branch interest very much to the fore. To ensure that the regional policy as a whole should become a matter of the first importance to ministers and their officials, the administrative structure and thereby the interests and priorities of its officials would have to be changed.

The Labour Market

The distribution of the labour force played a vital role in the implementation of regional policy during the five post-war years, a fact that was recognised by contemporary economists and policy-makers.[8] Some 20 million Soviet citizens had died during the war and the impact of the purges had left Soviet industry very short of labour even in 1940. Workers were therefore in extremely short supply in 1945. However, the demobilisation of some 8.5 million servicemen over the next three years and a continued migration from the land led to the number of workers and employees in the national economy attaining a 1950 level more than 25 per cent above that expected by the planners in 1946.

A substantial number of workers returned to the liberated areas of the west. The industrial labour force of these areas in 1950 was 7 per cent greater than in 1940, in spite of wartime losses of up to 80 per cent there. However, that meant that the proportion of USSR industrial personnel working in these areas was smaller in 1950 than in 1940 although well above its 1945 level. In Central Asia and Kazakhstan, in contrast, the relative gains of the war years were partially reversed in 1946–50, although in absolute terms the labour force did slightly increase in size over the post-war period, as Table 4.4 shows.

TABLE 4.4. Regional distribution of the Soviet labour force 1940, 1945, 1950

	Workers and employees in industry (at September) in 1000s					
	1940	*Per cent of USSR*	*1945*	*Per cent of USSR*	*1950*	*Per cent of USSR*
USSR	13079	100	10665	100	15317	100
RSFSR	9025	69	8076	75.7	10827	70.7
Siberia and Far East[a]	881.2	6.7	1230.7	11.5	1752.1	11.4
UKRAINE	2614	20	1256	11.8	2509	16.4
BELORUSSIA	394	3	152	1.4	346	2.3
MOLDAVIA	23	0.2	30	0.3	52	0.3
LITHUANIA	5.7	0.4	50	0.5	97	0.6
LATVIA	113	0.9	75	0.7	171	1.1
ESTONIA	73	0.6	55	0.5	106	0.7
GEORGIA	130	1	125	1.2	175	1.1
ARMENIA	44	0.3	45	0.4	81	0.5
AZERBAIDZHAN	139	1.1	124	1.2	173	1.1
UZBEKISTAN	182	1.4	239	2.2	254	1.7
KAZAKHSTAN	177	1.4	304	2.8	365	2.4
KIRGIZIA	36	0.3	55	0.5	66	0.4
TADZHIKISTAN	31	0.2	33	0.3	44	0.3
TURKMENISTAN	41	0.3	46	0.4	51	0.3

Source: See Note 9 on p. 159.

[a] Data are for average numbers of workers and employees over the year. The September 1940 figure for these areas was 894.2 and that for September 1950 1782.7 according to *Nar. Khoz. RSFSR* (1957) which differs from the above.

In Siberia and the Far East, however, the labour force continued to grow at roughly the national average rate after the war. The share of the Baltic republics and Moldavia in the All-Union industrial workforce rose rapidly in the Fourth Five-Year Plan period. In strictly quantitative terms, therefore, the eastern regions of the USSR were not particularly short of labour (although the Urals region *may* have been so), and the new areas of the west obviously benefited from changes in the distribution of labour over the plan period.

It is, however, by no means certain that the quality of the labour force was uniform in all regions. A Soviet economic historian has attributed the failure of West Siberia to fulfil its output plan in 1946 in part to the shortage of qualified workers.[10] The return of wartime evacuees to the west and of women and pensioners to the countryside led to a steady decline in the proportion of experienced and qualified (*kadrovye*) workers even in such a high-priority project as the Kuzbass coal combine over the period 1945–7.[11] Labour turnover was also abnormally high in the east during the first two years of the plan period. Experienced and qualified workers willing to stick to their jobs were scarce enough in other areas. Regions under reconstruction, like Leningrad and the Donbass, also complained of a lack of skilled workers and alarmingly rapid labour turnover. Even the Moscow coal basin, which had regained its pre-war output levels long before the end of the war, was still suffering in this respect.[12] All this was despite the draconian labour laws of 1939–40 which were supposed to restrict labour mobility. They were not effectively enforced after the war.

It is known that the industry of Siberia and the Far East received only 3 per cent of the Soviet servicemen demobilised after the war.[13] Some of this inequality in distribution might have been redressed by the training of 419,000 new workers for these areas over the five-year period in the schools of the Ministry of Labour Reserves and elsewhere. G. A. Dokuchaev claims that the cadres problem in Siberian industry was 'solved' by about 1948.

In spite of this there is some evidence that the east of the RSFSR continued to suffer from a lack of properly trained cadres right through the plan period. A decree of 30 August 1946 criticised the training schools for poor work, noting particularly their lack of proper teaching equipment and buildings and of trained teachers.[14] The result in the east was that all too often people were trained on the job by workers not much more experienced than their pupils.[15] Did such 'training' really provide better workers? In addition, the number of workers trained in the east fell off sharply during the last two years of the plan. Just before the 19th Party Congress in 1952, the Deputy Minister of Labour Reserves acknowledged that the east of the RSFSR lacked sufficient training schools.[16] However, his ministry could be accused of paying more attention to the needs of the west rather than the east; twice as many workers, for example, were trained by their schools in the Ukraine in 1947 and 1948 alone than in Siberia and the Far East during the entire plan period.[17] Similarly, the industrial branch ministries were quite capable of transferring existing workers from one plant in the east to another in the west.[18]

The training of the workforce to raise its quality was another matter from its recruitment. The decree of August 1946 noted above had criticised the Ministry of Labour Reserves for *re*training too many existing workers rather than training newcomers to industry. It was only in May 1947 that the Ministry was charged with responsibility for mass recruitment of labour on an individual basis. Even that reflected a system of recruitment that relied more on providing incentives to potential recruits than on the 'administrative methods' of recruitment practised during the war.[19] The effect of trying to lure young people, kolkhozniki, and so on, into industry and to particular areas, rather than directing them, was to allow the industrial-branch ministries more influence over the distribution of the trained labour force. The Ministry of Labour Reserves relied on these ministries for the supplies of housing, services and consumer goods that could give real force to the wage incentives of the 1946 and 1947 decrees.

In practice, however, it appears that these ministries could not even be relied upon to pay the centrally determined wage rates. Wide regional and inter-enterprise variations in wage levels were ended only by the 1956 reforms.[20] In spite of the decree of 25 August 1946 average wage levels of industrial and construction workers in Altai and Krasnoyar Krai in 1947 were still some 10 to 15 per cent *below* the RSFSR average.[21]

The monetary bonuses promised to workers willing to go east were not always backed up by supplies of housing, services and goods upon which to spend the extra wages. Housing construction was largely a matter for the All-Union branch and construction ministries; they were to build 65 of the 72.4 million square metres of new urban housing promised under the Fourth Five-Year Plan. Accommodation was in especially short supply in the east and these ministries were often attacked for failing to ease this shortage. In the chaos of 1945 several ministries had fulfilled only 35 to 40 per cent of their housing construction plans.[22] Dokuchaev cites the housing shortage as the main reason for high labour turnover in Siberia in 1946. The head of the Bashkir Oil Union in the Urals laid the blame for the lack of housing there on the Ministry of Fuel Enterprises Construction.[23] Delegates to a 'scientific-production' conference in Molotov (now Perm) oblast' criticised the lack of living accommodation as a major hindrance to the development of the local Kizelovsk coal basin as late as 1948. This appears to have been a general problem for the eastern coal industry.[24] Indeed *Izvestiya* reported that only in that year was the *first* centrally administered housing construction combine in the Urals region being constructed, and the development of that was being frustrated by delays in the Ministry of Heavy Industrial Enterprises Construction in Moscow.[25] The housing shortage was, however, by no means confined to the eastern areas. A Central Committee decree of January 1947, for example, noted it as among the reasons for Belorussia industry's poor performance in 1946.[26]

The same decree called for more construction and reconstruction of basic services, schools, hospitals and utilities in Belorussian towns to solve the cadres problem there. Where such services had been destroyed in the west, they had often never been built in sufficient quantities in the east. Replying to criticisms from the miners of Tula (with whom they were engaging in 'socialist competition') representatives of miners from Chelyabinsk in the Urals blamed their poor performance partly on the poor provision of social services. In the same issue of *Izvestiya* the lack of new schools in Irkutsk, eastern Siberia, was criticised. In January 1948 the Ministry of Fuel Industry Enterprises Construction was attacked for failing to build canteens, clubs, bathing facilities and the like for workers (and potential workers) at a new oil combine in the Fergan Valley in Uzbekistan.[27] Besides voicing complaints in newspapers and at conferences, local officials also had other channels for articulating their grievances at the All-Union level. For example, V. I. Nedosekin, the First Secretary of the Sverdlov *obkom* in the Urals, called for much greater expenditure to meet the 'cultural needs' of the population in his region at a 1947 session of the Supreme Soviet.[28]

The distribution of mass consumption goods was also vital to any effective policy on labour supply, and hence to regional policy. There had been a severe drought in many areas in 1946 and this had delayed the end of food rationing until late 1947. In the light of this it was no surprise that extra food and other supplies were amongst the bonuses promised to eastbound workers by the decree of May 1947. In spite of this the chairman of the RSFSR Gosplan noted the lack of such goods at a meeting of local planning officials in 1950.[29] The supply of foodstuffs and consumer goods was, however, only partly the responsibility of the supply departments of central ministries. Locally controlled industry must take some of the blame for the poor performance of Soviet light industry in these years. Even so the provision of mass consumption goods to non-eastern regions was by no means satisfactory. Disquiet as to the generally poor supplies 'to the regions' of such goods was voiced by no less an authority than Malenkov at the 19th Party Congress.[30]

In summary, therefore, eastern industry did suffer badly from a shortage of stable and trained industrial workers after the war. This shortage was in turn due in great part to the poor living conditions in these areas, which the bonuses offered by the government in 1946 and 1947 could not completely offset. It was perhaps significant that the main source of new labour in Siberia and the Far East were the local kolkhozy.[31] Even the ill-equipped Siberian towns might well seem attractive to collective farmers suffering from bad weather and extremely low procurement prices.

Other areas of the USSR, especially those liberated from Nazi occupation, also suffered from severe labour problems associated with poor living conditions. Their labour forces did, however, expand much more rapidly than those of the east (especially than those of the Central Asian republics). Only in the case of the 'new areas' of the west is there any evidence that this was due to central policy decisions.[32] The Soviet labour market was at this time a relatively free one. The regional distribution of labour was controlled not so much by central decree as by the incentives individual areas and organisations could offer. Those incentives depended greatly on the decisions taken (or not taken) in the branch and branch-construction ministries in Moscow. In fact, the general failure of the Ministry of Labour Reserves to control the distribution of the workforce often led these ministries to recruit their own labour on a scale that was frankly illegal.[33] The advocates of regional rather than branch interests could only seek to complain to central authority through press and conferences and thereby seek special consideration of their cases.

Material Supplies and Construction Work

The market for material inputs to construction trusts and industrial enterprises was, in theory, far more tightly controlled by the plan than that for labour. In practice, however, the inadequacy of supplies of everything from bricks to coal and machinery was one of the main reasons for the relatively poor performance of eastern industry during the Fourth Five-Year Plan period. This is reflected in the levels of labour productivity in these areas presented in Table 4.5.

This crude aggregate index of industrial labour productivity actually declined in Siberia and the Far East over the period 1945–50, whereas it more than doubled in the liberated republics. In Central Asia, Kazakhstan and the TransCaucasian republics it rose, but not to very far above its 1940 levels. These differences may have been partly due to variations in the quality of the labour force, but they can also be attributed to the poor supplies of capital and current inputs to eastern plants. The evidence for such a conclusion rests mainly on the mass of complaints voiced at the time by long-suffering officials in those regions. The target of these complaints was almost invariably the All-Union Ministry, its glavki (especially the supply glavk), or even the minister (or his deputy) himself.

TABLE 4.5. Labour productivity in Soviet industry by regions 1940, 1945, 1950

1940 = 100	*Industrial output per man (industrial output ÷ no. of industrial workers and employees at September)*	
	1945	*1950*
USSR	113	148
RSFSR	118	146
Siberia and Far East[a]	136	123
UKRAINE	54	120
BELORUSSIA	52	131
MOLDAVIA	34	91
LITHUANIA	46	112
LATVIA	71	200
ESTONIA	97	236
GEORGIA	83	116
ARMENIA	91	135
AZERBAIDZHAN	87	112
UZBEKISTAN	81	131
KAZAKHSTAN	80	113
KIRGIZIA	80	117
TADZHIKISTAN	70	106
TURKMENISTAN	78	115

Sources: As Tables 3.1, 4.1 and 4.4: see Note 34 on p. 160.

[a] Figures for Siberia and the Far East are based on different labour data (see Note to Table 4.4). Using comparable data the RSFSR figure for 1950 would be 132.

The frequency with which these particular targets were cited, added to the evidence of shortfalls on the supply side in the Soviet economy, suggests that the complaints were something more than a device used by local officials to divert attention away from their own managerial shortcomings. The supply of inputs was certainly the major constraint on a region's economic performance at this time, and in this sphere the ministries held the trump cards.

Such complaints abounded from the earliest years of the plan period. In August 1947, for example, *Izvestiya* quoted a letter it had received from Kazakh deputy to the Supreme Soviet. He attributed the slow construction of a new copper-ore extraction combine to

> a lack of attention to this project from the Ministry for Heavy Industrial Enterprises Construction and the Ministry of Non-Ferrous Metals, who are not ensuring the establishment of . . . mechanisms, equipment and materials. All work is being done by hand.[35]

This was by no means an uncommon method of construction at that time. A USSR Gosplan official in Kemerovo, West Siberia, made similar criticisms of his local construction trust, which was part of the same construction ministry.[36] The plans for new construction at the Gornaya Shoraya iron ore workings in 1946 had only been 60 to 70 per cent fulfilled; the ministry had promised to improve matters, but performance over the first six months of 1947 had shown little improvement. Gosplan's general concern at the slow rate of coal-mine construction in the east was spelt out in a leader in *Planovoe Khozyaistvo* in early 1947.[37] The Ministry of Fuel Industry Construction was singled out as the responsible party. The same ministry was criticised in the national press by the head of the Bashkir Oil Union.[38] The ministry's local trust had actually completed only 10 per cent of the work planned for it at this site in the first two quarters of 1947! The official channel for such a complaint would have been to the Oil Union's superiors in Moscow (the Eastern Areas Oil Ministry), from thence to the construction ministry in Moscow, and thence to the local construction trust. The head of the Oil Union was, however, exploiting his position as a deputy to the Supreme Soviet to air his plight in the national press in a bid to secure intervention from a higher level in Moscow.

Complaints against the local trusts of this ministry continued. In 1948, for example, another deputy accused the Central Asian Oil Industry Construction Trust of not fulfilling its plan obligations.[39] This ministry, established in 1946, was reincorporated into the Ministry of Heavy Industrial Enterprises Construction in December 1948. The reuniting of the regional coal and oil ministries no doubt lessened the need for its co-ordinating role, but its poor performance must have hastened its demise.

It was only in May 1950 that further leadership recognition was given to 'serious shortcomings in the direction of the construction ministries'.[40] The decree concerned criticised the spreading of construction effort over too many projects (so that relatively few were completed in time), the inefficiency or even total lack of project planning (leading to spiralling construction costs), the lack of mechanisation in the construction industry, and the shortage of construction materials. The solution was to establish a new State Committee on Construction (*Gosstroi*) to co-ordinate productive effort on this front. How successful this Committee and the ministries were in remedying these deficiencies can be judged by the complaints heard a year later at regional and All-Union Party Congresses. For example, the Kazakh Party Congress in September 1952 was told by its First Secretary, Shayakhmetov, that plans for the construction of both coal mining and oil-extraction enterprises were still being persistently underfulfilled.[41] The draft directives of the new Fifth Five-Year Plan called for the strengthening of the metal industry construction trusts of the Ministry of Heavy Industrial Enterprises Construction 'especially in the eastern regions'.[42] The approved version of the plan extended this demand to the coal industry and the railways, as well as the construction trusts of the branch ministries.

This recognition of the problems of new construction in the east came only after six or more years of complaints by local and Gosplan officials, Supreme Soviet deputies and the like. Yet as early as June 1947 the responsibility for this distortion of the planned regional policy had been correctly allotted by A. V. Korobov, head of Gosplan's Capital Construction Department and long-time supporter of complex development in the east. Citing examples from the Urals and the Central Asian coal industries, the Siberian iron-ore extraction and eastern automobile construction sectors, and the eastern textile industry, he concluded that

> In the past year there have been serious shortcomings in the fulfillment of the capital work plan in the eastern regions of the USSR. Many ministries and organisations paid insufficient attention to the business of the development of the eastern regions, and to the creation in these regions of new production bases, which are necessary to speed up the growth of production and to liquidate irrational and overlong transports.[43]

However, the blame cannot be laid only at the door of the construction ministries and the ministries supplying machinery, or even of the construction trusts of branch ministries. For a major reason for slow construction work in the east was the chronic shortage of construction materials—bricks, cement, timber, glass, steel and so on. Their production and distribution was the responsibility of the Ministry of Construction Materials in Moscow, as well as of other branch ministries and local Soviets.

The need to reconstruct in the west and to continue to expand construction in the east meant that all construction materials were bound to be in very short supply over the post-war years. This bottleneck in the economy was to be eased partly by the appointment of Kaganovich as Minister of Construction Materials over the period 1944–7. However, not even Kaganovich's often coercive 'trouble-shooting' style, which he applied to both the railways and the Ukrainian party over this period, could seriously affect the shortage of building materials in the east.

For example, *Pravda* reported in May 1946 that a new fertiliser plant in Molotov oblast' was working at only 45 per cent capacity, as a shortage of construction materials was preventing the carrying out of essential capital repairs. The Ministry (of the Chemical Industry) had responded to previous complaints by simply ordering the plant to get the repairs carried out by August, without providing the means to do so. Indeed, the supply glavk of the ministry had not even issued all the appropriate orders for metal parts to the Ministry of Iron and Steel, let alone exerted any pressure to secure the fulfilment of the orders. The ministry (under Pervukhin, later a Politburo member) was attacked for generally ignoring the needs of the factory. In addition, *Pravda* revealed two days later that the whole of Molotov oblast' was short of bricks, the local brick plant lacking supplies of fuel, electric power and labour.[44] In spite of continuing protests over the next five or so years, F. M. Prass, by then First Secretary of the Molotov obkom, revealed at the 19th Congress in 1952 that 'the production of building materials is a bottleneck that is holding up capital construction in our oblast''. This he blamed particularly on the Ministry of Construction Materials and its head, P. A. Yudin, noting especially the latter's failure to expand brickyard capacity in the oblast'.[45]

This shortage of materials was by no means confined to one oblast' in the Urals. A decree of December 1946 noted that shortages of building materials were hampering the work of the Far Eastern Construction Trust *Dal'stroi*.[46] *Izvestiya* reported in mid-1947 that Novosibirsk oblast' was short of bricks and timber; indeed brick production plans for this important oblast' were being only 20 per cent fulfilled.[47] Delegates from several eastern regions of the RSFSR at a 1950 conference of local planning officials criticised enterprises of the Ministry of Construction Materials for not producing at anything like full capacity.[48] Earlier in the year the decree on lowering construction costs had blamed this ministry and the Iron and Steel Ministry for the slow rate of expansion of building materials output. The Minister for Construction Materials, S. Z. Ginzburg, who had headed various construction ministries since 1939, was dismissed, although his replacement, P. A. Yudin, had himself been the target of much criticism as Minister for Heavy Industrial Enterprises Construction. Perhaps more significantly from the regional point of view, this decree made particular reference to the need to expand building materials output in the Urals, Siberia and the Far East, partly to remove the need to transport such materials from the west.[49] With the exception of the establishing of Gosstroi, the only change in administrative structure promulgated in this decree was the transfer of many outside construction materials enterprises to the control of the much-criticised ministry. Inadequate supplies of building materials therefore hindered the construction of new enterprises and the expansion and repair of existing plant in the east.

Eastern enterprises also suffered from uncertain and tardy supplies of new machinery. Officials of a Kizelovsk coal combine attributed their poor output record to this in 1947, although *Izvestiya* rejected their claims.[50] A party organiser at a wagon-manufacturing plant in Chelyabinsk oblast' reported in September 1947 that production of tramway coaches had not yet even begun there, as the Dinamo plant in Moscow had not supplied vital equipment. In spite of an April order from the All-Union Council of Ministers to ensure to the plant all necessary supplies, the Minister of the Electrical Industry still could not guarantee their delivery within the 'forseeable future'![51] Even such high-level intervention could not always secure supplies.

Poor supplies of current inputs (fuel, raw materials, and so on) as well as capital inputs (new construction and installation of machinery) held back the development of the eastern industrial bases. The Tomsk obkom archives for 1946, for example, are full of complaints about inadequate supplies. The difficulty of securing supplies in winter meant that the Kuzbass mines were working at only 50 per cent capacity in 1946–7.[52] Indeed one economist claimed in 1948 that

> Quite often the machinery established at an enterprise is only 50 to 60 per cent utilised, and that is a consequence of uncompleted machinery, poor production planning, and unsatisfactory organisation of material supply.

A reviewer commented that such figures were not credible[53] but, in the light of the evidence so far presented, one might conclude that they are only too credible, especially in the eastern areas.

The organisation of material supplies may have been improved by the establishment of a State Committee on Supply (*Gossnab*) in December 1947. Before that material supplies had been largely a matter for the supply departments (glavki) of branch ministries. These glavki were renowned for fulfilling only those contracts they wished to fulfil, and, especially, for favouring their own ministry's enterprises. As a result many branch ministries became more like separate empires, building their own houses, producing many of their own materials, and even building their own machinery. Co-ordination of productive activity was primarily within branch not regional boundaries. This attempt to restore co-ordination along functional lines (with separate construction and machine-building ministries, as well as Gossnab) might have lessened the dominance of branch interests. However, the decree splitting off Gossnab from Gosplan only called for 'preparatory work' towards the incorporating of *some* ministerial supply glavki into the new committee.[54] These glavki continued to distribute scarce supplies as long as the ministries existed; Gossnab could only intervene in selected cases. Like the officials of Gosplan it could issue paper orders and make examples of particular cases, but the ministry still made most of the day-to-day decisions as to what was to be supplied to whom.

Transport

One grave shortcoming in the work of ministerial supply departments was that they worked out their plans without taking into account differentials in transport costs. Until 1949 wholesale prices did not include any allowance for the cost of transporting the produce. Supply glavki, therefore, had no interest in cutting out overlong transports of goods. Only with the wholesale price and freight tariff reforms of 1949 were these glavki 'obliged to make a systematic study of the geographic distribution of production and consumption, to draw up regional balances of production and consumption . . .'.[55] It was due in no small part to their failure to do so before that date that, as Malenkov noted at the 19th Party Congress: 'Irrational and excessively long railway hauls have not yet been eliminated'.[56] The complex development of each economic region had been intended to reduce inter-regional transports; in practice the amount of freight carried between regions by the Soviet railway system rose by 62 per cent between 1946 and 1950.[57]

The implementation of complex development policies was also being frustrated in the transport sector by the activities of the Ministry of Transport itself. The Fourth Five-Year Plan had called for the construction of 7230 km of new main line on the railways, 35 per cent of which was to be laid in the Urals and Siberia. Yet of this only 2319 km were actually built. Two-thirds of the capital resources expended on the railways went to the liberated areas. The effect of this on the east may be illustrated by two examples. First, in 1947 Gosplan's representative in Kemerovo oblast' complained that the Ministry of Transport was refusing to issue any directive concerning a planned railway link from Abakan to Stalinsk (now

Novokuznetsk) to permit the exploitation of iron ore deposits at Abakan. The Ministry would not even say who was to build this 400 km line or when. Similarly, I. R. Razzakov, the First Secretary of the Kirgiz Party at the 19th Congress in 1952, accused the ministry of delaying the construction of new railway links for the Uzgen coal basin.[58] Again, the development of a new industrial complex in the east was being hindered by the actions of a branch ministry that had little interest in implementing any regional policy.

Neither were plans for technical improvements such as electrification and the introduction of block-signalling systems fulfilled during the post-war years. Overall plan targets were met by forcing up norms for the utilisation of existing equipment and introducing strict party discipline on the railways under the political department (*Politodel*) system. However, the Urals and Siberian industry's demands for transport facilities had greatly exceeded their supply even in 1945.[59] As a result the transport bottleneck in these areas (especially on the Trans-Siberian route) became even more acute after the war. Not even iron discipline could get many more trains on to one track at one time!

Typical of the complaints about poor transportation work on the railways in the east was that of two economists in 1947, who claimed that much of the coke produced in the east of the RSFSR was being ruined by transportation and storage problems long before it reached the blast furnaces of the Urals and West Siberia.[60] The head of the Bashkir Oil Union who had previously complained of shortcomings in the work of other ministries also attacked the Transport Minister, Kovalev, for failing to provide enough oil tank wagons to transport the output of his refineries. Delegates to the scientific-production conference in Molotov oblast' in April 1948 also criticised transport bottlenecks in the Kizelovsk area.[61]

The Urals and Siberia suffered not only a shortage of transport links with other regions, but also from a shortage of track both within factories and linking factories to the main line. The Five-Year Plan had called for improvements in this respect, but complaints of poor loading and unloading work in factories and a shortage of spurs to factories in these areas continued.[62]

A leadership's response to these problems was to revise the tariffs for haulage of railway freight in 1949 and again in 1950. As early as 1946 the economist Turetskii had argued that

> the improvements in the distribution of productive forces in the new five year plan, and the shortening of the length of transport hauls also

demands the introduction of a series of correctives in the system of transport tariffs.[63]

The old tariffs had been set in 1939 at the time of the adoption of policies of complex development and had been set to discourage overly long transport hauls. The new tariffs were generally higher than their predecessors and were supposed to reflect the actual cost of transport in most cases. However, the tariffs for extremely long journeys were set above the level of costs to further discourage them. Some especially low tariffs for high priority projects (such as the Kuzbass-Magnitogork link) were also raised slightly. At the same time prices for extremely short journeys (less than 50 km) were also raised to drive such hauls on to the roads. The aim of the new tariffs was clearly to facilitate complex development,[64] but with the revision of many of the new industrial prices, freight rates were also cut. In January and July 1950 tariffs were reduced and with them the incentive to cut transport hauls and further complex development. It was argued that this would cut the cost of construction,[65] but would have done so in monetary rather than real terms.

For most of the first post-war plan period railway construction proved a major stumbling block to the further development of eastern industry and the Ministry of Transport did little to ease the problem either in its construction record, or its provision of transport services, or in its price-fixing activity. Other forms of transport fared little better; for example, Saburov in his report on the new plan directives to the 19th Congress revealed a lack of transportation facilities on Siberian waterways.[66]

Administration

The relative failure of eastern industrial expansion may also have been due in part to deficiencies in the organisation and administration of industry in those regions. For example, Siberian industry was short of experienced and qualified administrators and specialists. Less than 10 per cent of the 'directing cadres' in Siberia and the Far East in November 1947 had received higher education; 38.1 per cent had less than three years experience in their occupations.[67] In addition some organisations in the east were publicly criticised for their handling of economic problems, notably the Kemerovo obkom in 1948.

More basically, however, the entire administrative machine in the Soviet Union seemed to have a built-in bias against the eastern regions and, indeed, against any effective planning of the regional distribution of industry. The leadership's regional policy had to be implemented by ministries organised along functional and branch rather than regional lines. These ministries were responsible in effect for deciding where new construction projects were to be completed, to which regions labour would be attracted, and whence supplies of inputs of all kinds were to be directed.[68] They directed these towards the areas that yielded most rapid returns in a bid to maximise branch rather than regional output.

It is true that most ministries did have regional glavki and at least three ministries—those of oil, coal and fisheries—were divided into separate regional ministries in 1946. This however, did nothing to ease the problem of supplies from other ministries and therefore of co-ordination of productive effort within economic regions. Indeed it raised new problems of co-ordination within the branch without really solving those of co-ordination within regions.[69] As a result of these ministries were reformed as single production-branch ministries in 1948–9.

Bodies did exist in the vast Soviet administrative complex that might have provided such co-ordination. Party, Soviet and state planning and financial organs could have represented the interests of individual regions and of regional policy aims against the branch-oriented activities of ministries. In practice, however, they lacked the power and sometimes the will to provide an effective check on ministerial authority.

It has been noted by a more recent Soviet student of this period that the system of 'vertical' (branch ministerial) administration of the economy prevented any effective planning for the complex development of particular regions and in the process 'violated the rights' of local party, Soviet and economic organs.[70] Even during the post-war years themselves, the economic geographer Kolosovskii claimed that the experience of the three previous Five-Year Plans showed the grave difficulties inherent in trying to co-ordinate the activities of two neighbouring enterprises if they were subordinated to different *vedomstva* in Moscow.[71] The well-known economic commentator A. A. Arakelyan went so far as to argue that 'The task of further expanding the economic power of the union republics can be successfully realised by means of enhancing the role of the republican organs of economic administration.'[72] This he reiterated in 1947, emphasising also that the individual enterprise should be freed from 'petty and unnecessary supervision' by the All-Union ministries.[73] It was well into 1950 when a local Soviet official from Irkutsk bemoaned the lack of effective complex planning of individual economic regions.[74] Indeed the basic statistics and analyses, upon which both that planning and any attempt to check on regional policy fulfilment had to be based, were in sadly short supply.[75]

Indirect Controls over Regional Policy: The Checking Organs

It was therefore officials of production ministries based on branches of industry who were largely responsible for the distortion of regional policy. As branch officials they had little interest in carrying out a policy of regional balance; of the checking officials within and outside the Council of Ministers structure some had similar branch interests. Others did have responsibilities for checking on policy execution within an area. Yet these often lacked the authority and the resources to counteract the power of the ministries.

The Party

We have already outlined the weaknesses of party *Kontrol* in Chapter 2. The party's full-time apparatus lacked numbers and expertise and were divorced from their rank and file membership who could have helped to check on policy implementation at grass-roots level.

In the sphere of regional policy the central apparatus in Moscow lacked any interest in its fulfilment. The Central Committee Secretariat was organised along functional and then (from 1948) functional and production branch lines. Its staff had responsibility for checking on the fulfilment of plans by branch and function rather than by region. In so far as the central party apparatus structure paralleled that of the ministries its interests did not differ from the latter's.

The party did of course have an extensive regional network. Would not republican and oblast party secretaries and inspectors show a keen interest in 'pulling up' ministries who were diverting resources to other areas? It was, however, at these and lower levels of the party that the shortage of qualified cadres was especially acute. Secretaries of regional party organisations often criticised their subordinates for poor control over plan implementation at the grass-roots level. For example, Khrushchev made this point in the Ukraine in 1948 and Shayakhmetov for Kazakhstan in 1952.[76] Indeed several local party organs were criticised at the 19th Party Congress for acting as expediters of supplies for local factories rather than checkers on policy implementation. Regional party officials from the east could complain about ministerial indifference; in practice they could do little more than put local managers in touch with alternative sources of supply. It was, after all, the central party apparatus, and, after 1948, its branch sections, that had the power of appointment over ministerial positions. Those parts of the party that had an interest in checking ministerial power thus lacked the resources and the powers to do so effectively.

Checking Agencies within the Council of Ministers

One might have expected All-Union planning officials to have taken some interest in ensuring correct regional proportions. The complaints against ministries by Gosplan officials like Korobov shows that at least some of them took this point of view, but we also know that other planners had favoured short-run specialisation as against complex development policies in 1945. As a result Gosplan as a body seemed only half-committed to complex development policies. A delegate to the 19th Congress claimed that 'In the USSR Gosplan the department-oriented claims of individual ministries are still being supported by certain officials and the interests of the state are thereby being violated'.[77]

Above all, however, Gosplan lacked the teeth to ensure the fulfilment of the regional policy of 1946. If its All-Union officials were disunited, Gosplan's regional officials were simply too thin on the ground. No doubt Gosplan's Regional Distribution Department had by the 1940s acquired rather more than the twenty-three personnel it had had in 1936, but its major weakness seemed to be in its links with individual enterprises and areas. The sectors for each union republic within this department were linked to individual krai and oblasti by representatives (*upolnomnchennye*), who were supposed to exercise the 'day-by-day' kontrol over the enterprises in their areas. Their crucial role in checking on the fulfilment of regional policy was emphasised again and again in Gosplan's journal.[78] They were charged with ensuring co-operation amongst different industrial branches within their area, and between that area and other areas. They were to assess the potential resources of their area and keep records of material balances there. The representatives also had to report on plan fulfilment in his oblast' in aggregate terms and for each branch of industry. Yet these officials seem to have had, at best, only five or six assistants; the total Gosplan apparati in each economic region in 1940 had been only thirty to forty officials.[79] They could scarcely check in any detail on the work of all the enterprises in one oblast'. For example, the sheer paperwork involved in supplying reports on plan fulfilment to Gosplan, to All-Union ministries, and to local party organs must have taxed them to the limit. In a bid to lessen the burden on them and on Gosplan in Moscow, a decree of December 1947 devolved the functions of gathering statistics and of assessing material balances on to the newly created Central Statistical

Administration (TsSU) and Gossnab respectively. The effect of this, however, was to isolate Gosplan even further from day-to-day control over industry. Thus the Kazakh First Secretary could accuse Gosplan in 1949 of ignorance of local conditions and of a 'formalistic' attitude towards the problems of the regions.[80] As to the checking activities of the TsSU, its role was confined to reporting the non-fulfilment of regional plans. Corrective measures could only be taken by its superior, the Council of Ministers, through other organs, notably, of course, the ministries and Gosplan! In fact no detailed regional statistics were made public until the publication of the first reasonably systematic statistical handbooks in 1956–7. The impact of Gossnab on regional policy fulfilment has already been doubted. It is perhaps sufficient commentary on its performance that it was reincorporated into Gosplan within days of Stalin's death in 1953.

Little is known of the organisational structure of the secret police at this time. One half of it, the MVD, ran its own economic empire based on prison camps in the north and east of the USSR. The impact of that on the regional distribution of industry is virtually impossible to assess. The camps were located in the most underdeveloped areas of the Soviet Union but how far was their output reflected in official statistics? In view of the fact that some Western observers have pointed to the unproductive nature of such forced labour,[81] perhaps we should regard the MVD's primary aim as being to keep its prisoners away from the rest of society rather than to check on the implementation of regional policy.

The MGB was organised along regional lines. However, its main task at this time was not the sniffing out of economic 'saboteurs'. Far more stress was laid on dealing with 'bourgeois nationalist' survivals in the new and liberated areas of the USSR and on counteracting the lax political discipline of the war years. In any case, as we have outlined in Chapter 2, the power of the police in all spheres was by no means as great after the war as before it. This is especially true of the economic sphere.

The financial organs of the Council of Ministers headed by the Ministry of Finance were more directly economic agencies. Their checking role was being encouraged by the leadership, especially from 1948 with campaigns for full financial accountability (*khozraschet*) and kontrol by the rouble' in industry. The impact of these campaigns was marginal, however. Financial responsibility and accurate book-keeping were surely the exception rather than the rule in the command economy. Furthermore, neither the Ministry of Finance nor its subordinate banks had any particular interest in using financial levers to the benefit of *regional* policy implementation. The central apparatus of Prombank, for example, was organised basically along production-branch lines, with some functional departments.[82] As with so many other checking organs only its central apparatus could have had the power to counterbalance the branch ministries and yet it was so organised as to share common interests with them. Only its local organisations had a regional interest and grass-roots knowledge; they lacked both human resources and political weight at All-Union level.

The discussion so far assumes that investment funds and real resources were directed by central officials. Yet some industrial investment was financed by firms themselves. Stricter financial discipline might have ensured higher profits and more reinvestment from them. Firms in the east might have been able to provide their own funds if they were not receiving them from above. On the other hand it must be remembered that finance alone was not enough; it was all too often real resources on which to spend the money that were lacking.

The Soviet government did raise the whosesale prices of a range of heavy industrial products in 1948. Now for the first time a number of firms had some prospects of earning a profit. Before this prices for their output had been those fixed in 1926–7 or by branch ministries themselves (for new products). By the 1940s, these were so unrelated to costs as to rule out profit-making and ensure total reliance on centrally directed investment funds. In spite of the 1948 reforms, in 1950 firms still financed only about 20 per cent of their own investments. The impact of these price changes on regional policy was limited by their partial reversal in 1949–50[83] and the fact that initially they made no allowance for transport costs. The organs of financial control were in no position to affect regional policy even by forcing firms to rely on their own resources.

Other Checking Agencies

The only such bodies that might have influenced regional policy were the local Soviets, but only on the same 'self-help' principle as envisaged in the price reform. Local Soviets lacked even formal powers of supervision over the activities of industrial enterprises controlled by the ministries in Moscow. The chairmen of local planning commissions could and did complain that in many areas All-Union enterprises were not working well and that too much of a burden was falling on local industry and construction,[84] but they had little authority to take action on these complaints.

Local Soviets were, however, responsible for 'local' and co-operative industry within certain centrally determined parameters (targets for output of and investment in local and co-operative industry in each republic were set in the Fourth Five-Year Plan, for example). Industrial enterprises under the control of republican Councils of Ministers and lesser regional bodies were usually fairly small-scale, labour-intensive, and concentrated in the light industry sector, although local initiative in developing small fuel deposits and construction-materials plants was at this time being encouraged. According to the Five-Year Plan local and co-operative enterprises were to produce 27.5 per cent of USSR industrial output in 1950 and local Soviets were to be responsible for the construction of 10 per cent of the new urban housing over the plan period.

Development of local industry was seen as an important source of complex development for the eastern regions. The output of local and co-operative industry in the east of the RSFSR was planned to at least double its pre-war level by 1950, which probably meant more than doubling its 1945 level.[85] However, 60 per cent of local and co-operative industry in the RSFSR (according to the plan) was controlled by the RSFSR Council of Ministers in Moscow. RSFSR ministries had no more interest in fostering development in the east than did their All-Union counterparts. The local Soviets of the Urals, Siberia and the Far East controlled probably only 3 or 4 per cent of Soviet industrial output.

Similarly in many other republics some 50 to 80 per cent of industrial output was planned to be produced by local industry, but in many cases this must have included industry subordinated to republican ministries who were in turn subordinate to All-Union ministries in Moscow (the All-Union Ministries of Light Industry, Textile Industry and Construction Materials were all so organized). The industrial enterprises controlled by local Soviets and small co-operative (rather than republican Councils of Ministers and therefore often an All-Union ministry) were to produce only 8 per cent of All-Union output in 1950.

In aggregate, local and co-operative industry did manage to fulfil all its annual plan targets and its output in 1950 was 50 per cent above its 1940 level. The resources of this sector of industry were, however, still not sufficient to counterbalance the anti-eastern bias in the administration of All-Union industry. Capital construction work for local Soviets was often very poorly executed and the low quality of local industry's output was frequently criticised.[86] The lag in local production of construction materials was criticised in a *Pravda* editorial in 1946.[87] Such criticisms were still being voiced four years later, especially with reference to Central Asia and to the east of the RSFSR.[88] The local Soviets outside the RSFSR had neither the power to check the ministries nor the resources to counteract their anti-eastern bias.

Conclusion: Responsibility for Regional Policy Outcomes

None of the controlling agencies discussed provided an effective channel of information and control through which the leadership in Moscow could co-ordinate the activities of individual ministries within each region. As a result decisions on regional distribution were taken (in practice) by ministerial officials on a purely *ad hoc* basis without reference to the regional policy promulgated in 1946. The economist R. S. Livshits wrote in 1954:

> In the post-war period in several cases the distribution of industrial enterprises and the choice of sites for their construction has not met the needs of the law of planned and proportional development of the national economy and has exhibited a series of shortcomings.[89]

The implementation of the location decisions that made up regional policy was left to branch and functional ministries who generally favoured the more developed industrial areas in their direction of both capital and current inputs. The party and state leadership in Moscow simply lacked an administrative machine that could rival the power of ministries over decisions at the enterprise level, and that had a material and political interest in enforcing the criteria envisaged by the Fourth Five-Year Plan. The leadership could, of course, intervene in specific cases of special concern, such as the Far Eastern fishing industry in October 1948 or the Volga-Don canal and irrigation project announced in December 1950, and the new areas of the west.[90] The interested apparatus needed to secure a nationwide regional policy was, however, lacking.

This is not to deny that in some respect the eastern regions of the RSFSR did meet their industrial plan targets. Most notably the plans for ferrous metals output on the 'eastern regions' (including the Volga region) were overfulfilled. However, it can be argued that in the production of iron and steel the Urals and especially Siberia were relatively developed, and thus could show rapid returns on investment for enterprises under ministerial control. Construction of the Urals-Kuznetsk combine had been begun in 1930, and these 'eastern regions' were already producing 30 per cent of the USSR's steel and pig iron in 1940. That this proportion was raised to a half by 1950 was not solely due to the problems of reconstructing the southern metal industry, but also to the fact that as early as 1940 productivity indices were generally higher in Siberia than in the Ukraine.[91] If the Ministry of the Iron and Steel Industry were seeking to maximise short-run returns it might well, therefore, tend to direct resources to Siberia.

The poor performance of eastern industry cannot be ascribed to shortfalls in the supposedly low-priority light industrial sector. The output of machine-building and metal-working plants in Siberia and the Far East rose by only 9.2 per cent over the period 1945–50, and these had accounted for 55.4 per cent of Siberian and Far Eastern industrial output in 1945. The output of the foodstuffs industry of these areas (9.9 per cent of total output in 1945) grew, in contrast, by 63.7 per cent over the five-year period after the end of the war.[92]

The problems of the machine-building sector may have been due to problems of reconverting to the production of civilian output in 1946–7. Yet Soviet specialists have always insisted that the reconversion process was completed by (at the very latest) 1947. One might well conclude that the resources needed for reconversion were often diverted elsewhere and/or that new construction in Siberia after 1947 was kept especially short of essential supplies. The ministries in Moscow controlled those supplies and none of the regional officials of the command economy had the power to stop them frustrating the leadership's policy.

An alternative explanation of these patterns of policy implementation is provided by the modified organisational model in the form of its emphasis on priority sectors. The analysis suggested by this model is that the leadership had always placed a higher priority on expanding industrial output as fast as possible (short-run maximisation) than on its regional policy of 1946, and thus that the policy outcomes of 1950 were basically as had been intended (although such lower priorities as the development of the east were not carried out as intended). This explanation relies, however, on the assumption that either the regional policy of 1946 was never intended to be fulfilled, or the leadership did not realise that it could not in practice fulfil both that policy and its aim of 'production at all costs'. The argument that the Fourth Five-Year Plan's provisions on regional policy were no more than window-dressing has already been considered. The general nature of those plans, the publicity accorded to the Fourth Five-Year Plan and subsequent decrees reinforcing its regional provisions, as well as the debates preceding it, cast doubt on that assumption. The nature of these debates and of leadership involvement in them, and the very fact that long-run maximisation criteria were built into the Five-Year Plan directives in 1946, argues against the assumption that the leadership did not realise that the two policy aims would clash. The plan embodied a compromise between those aims; its fulfilment tilted that compromise much further in the direction of short-run maximisation and the western areas. The best explanation remains that the leadership had two priorities and hoped to realise them both to a degree, but subsequently found it impossible to enforce one of them in the face of the power and vested interests of its branch ministries. The leadership was surely aware of this from all the complaints placed before it. It was not until the Khrushchev era, however, that any action was taken to strike at the root of the regional policy problem; the root was that the ministries had no interest in the enforcement of the policy of 1939 and 1946, and without such motivation did not feel compelled to implement it. The solution in 1957 was to place operational control over 70 per cent of industry in the hands of regional economic councils (*Sovnarkhozy*). Almost inevitably Sovnarkhoz officials were later criticised for 'localism' (*mestnichestvo*).[93] In the 'command economy' the leadership always had to pay due attention to the interests of the implementors be they regional or departmental. If it did not, its policies would not be carried out.

5 The Formation of Sectoral Policy 1945–53

Introduction

Our second case study of the operation of the command economy in the post-war years concerns the sphere of sectoral policy; that is, the balance of growth between different sectors of the Soviet economy. Reasons of space and data comparability must confine us to industry. The non-industrial sectors like agriculture, construction and transport are only of relevance in so far as they affected the balance within industry between its heavy industrial and consumer goods branches. The most convenient indicator of the relative priority of heavy and light industry in the USSR is that provided by the balance between the 'A' and 'B' sectors.[1]

The importance of this aspect of policy over the period 1917–41 scarcely needs demonstrating. The issue of how far industrial growth should be based on producing means of production (heavy industry) rather than consumption goods (light industry) was at the centre of the debates over the transition from war communism to NEP. It was one of the central themes of the industrialisation debates of the mid- and late-1920s.

The criteria upon which decisions on sectoral priorities might be based are very similar to those affecting regional policy. They may be summarised under the headings of short- and long-term maximisation of returns on investments, defence, and consumer satisfaction.

If the Soviet leadership sought to maximise short-run returns on its investments, then light industry, with its (generally) lower capital-output ratios and consequent shorter gestation periods was likely to yield more than heavy industry. A newly planned light industrial plant such as a textile mill can be expected to yield finished output and profits much more rapidly than a capital-intensive steelworks that might take many years to go into production. In so far as capital rather than labour was the scarcer resource in backward Russia, the initial construction costs of the steelworks would also tend to be greater.

Such arguments in favour of light industry are couched in terms of short-run maximisation of private returns to the investor. They formed part of the basis for the policy of NEP after the Civil War. Rapid returns on investments were needed to repair the damage done to industry by seven years of war.

In the longer run the greater initial construction costs of heavy industrial plants would be spread more thinly over the years. In addition, of course, the social returns from investment in heavy industry are much greater than from investment in light industry. A heavy industrial enterprise produces means of production which form the basis for the construction of further productive units. A light industrial enterprise produces means of consumption, which contribute relatively little to the further expansion of industry. Whereas steel can be used to build machines, and machines to produce other vital producers' goods, clothing can only be worn and then thrown away. Naturally clothing manufacture does have multiplier effects, but they are considerably less in the long run than those resulting from steel production. Long-run maximisation of all economic returns on investment (i.e. of economic growth) therefore demands more emphasis on investment in heavy industry than does short-run maximisation of private returns to the investor.

This criterion was the foundation for the early Five-Year Plans and the industrialisation drive from 1927–8. Heavy industrial growth was financed by diverting resources from agriculture and light industry. The share of the 'A' sector in Soviet industrial output rose from 39.5 per cent in 1928 to 61.2 in 1940.[2]

The survival of the Soviet regime depended on its ability to defend itself from its external (and its internal) enemies. To defend itself against the capitalist world the new Soviet state had to expand its production of weapons and ammunition, something that required machinery and the materials that only the heavy and extractive industries could provide. The survival of the Soviet state therefore depended on a certain level of investment in the defence industries and that in turn depended on a certain level of investment in heavy industry. Indeed it presupposed investment in heavy industry as most branches of 'war industry' are included in the 'A' sector of Soviet industrial statistics.

The defence criterion could be seen in the forefront of sectoral policy during the period of war communism (although some saw in this policy a longer-term attempt to build up the industrial base)[3] and in the pre-war Five-Year Plans. The defence argument for concentration on heavy industry was succinctly made by Stalin in 1931: 'We are fifty or a hundred years behind the advanced countries. We must make good this distance in ten years. Either we do it or they crush us.'[4] This argument became ever more pressing with the rise of fascism during the 1930s.

Clearly a certain proportion of industrial investment has to go to light industry to feed and clothe the population. The arguments over sectoral policy centred, and still centre, on how far consumer demands above and beyond this subsistence level should be satisfied at the expense of long-run industrial growth. It has often been argued that political stability requires a certain minimal level of satisfaction of consumer demands.[5] The survival of the Soviet regime even under Stalin required that they pay some attention to the demands and expectations of consumers. This was especially true of areas where the loyalty of the population was in some doubt, as in the case of the peasantry under NEP. Their demand for consumer goods had to be satisfied if they were to be persuaded to market the grain to feed the towns.

The Third Five-Year Plan (1938–42), however, emphasised long-run maximisation and defence considerations in its sectoral policy. The output of the 'A' sector was to grow by 107 per cent over the period 1938–42 and the 'B' sector by 72 per cent. The immediate pre-war sectoral policy was neatly summarised by V. Dyachenko in 1948:

> the basic share of the investments ploughed into other branches of the economy is realised by the gross revenue of these [light and food] branches of industry and of the procurements organisations. Thus in 1939 the share of the output of enterprises of food, meat and milk, light and textile industries and procurements in gross output equalled 36.8 per cent, whereas [their contribution to] revenue from turnover taxes equalled 87 per cent.[6]

The Impact of the Second World War on Soviet Sectoral Policy

The emergencies of war and the Nazi invasion necessitated a further concentration of resources on heavy industry from June 1941. Heavy industry's share of the total of Soviet industrial output rose from 61 per cent in 1940 to 84.4 per cent in 1942, much of the latter going directly to the war effort, as Table 5.1 shows. The disloyalty to the regime that might have resulted from a fall in personal consumption levels of 35–40 per cent[8] was partially avoided by the appeals to patriotism and defence of the motherland that permeated the war period.

As the German advance was halted and then turned back (from 1942–3), Soviet industry was gradually reconstructed along civilian lines and the

TABLE 5.1. Distribution of output by sectors of Soviet industry 1940–5

Year	Sector (per cent of gross industrial output) 'A'	'V'[a]	'B'
1940	61.2		38.8
1942	20.5	63.9	15.6
1943	22.1	58.3	19.6
1944	28.1	51.3	20.6
1945	34.5	40.1	25.4

Sources: See Note 7 on p. 162.

[a] the 'V' sector denotes 'war industries'.

share of the consumer goods ('B') sector in industrial output rose from a low of 15.6 per cent in 1942 to 25.4 in 1945. This was, however, some way short of its 1940 share of 38.8 per cent, which reflected the pre-war policy of priority to heavy industry. In other words, the war moved the scales of Soviet industrial growth substantially in favour of heavy industry for reasons which we will now detail.

The basic production capacity of Soviet light industry was concentrated in the Central and North-Western economic regions of the RSFSR[9] both of which were partially occupied by German forces. The losses of light industry were therefore substantial: 2.3 billion roubles for 'pure' light industry (i.e. 'B' sector excluding foodstuffs) and 22 billion roubles for the foodstuffs industry;[10] the total losses to state enterprises were 287 billion roubles. However the losses of heavy industry were also substantial: some 13.3 billion roubles in the coal industry alone.[11] One can only conclude that substantial losses were experienced by all sectors of industry due to wartime destruction. It is clear, however, that far more effort was made to evacuate and to build and rebuild heavy industrial than light industrial plant during the war.

Of the 1523 industrial enterprises evacuated eastwards in 1941–2 1360 were 'large war production enterprises'[12] and presumably, therefore, part of the heavy industrial sector. Only 350 light industrial 'plants' (sub-divisions of 'enterprises') were evacuated during the entire war.[13] First and almost overriding priority in evacuation was, not surprisingly, given to heavy industrial enterprises directly linked to the war effort.

Little attempt was made to build up new light industrial plant during the first half of the war before the start of mass reconstruction (August 1943). Investment in the light industry sector throughout the war years was only 60 per cent of its pre-war levels.[14] Again the lowness of those pre-war levels must be emphasised. The share of the pure light, food and other industrial sectors in all Soviet capital investment fell from 39 per cent in 1940 to 27 in 1942, and to less than 25 per cent in 1943.[15] The result of losses and lack of new construction in the 'B' sector was strict rationing of most consumer goods and foodstuffs by 1942.

After 1943 the 'B' sector did regain some of the ground it had lost. This was mainly due to the reconstruction of light industrial enterprises in the liberated areas of the USSR, However, much of this reconstruction appears to have been paid for by the efforts of workers and officials of light industry themselves. The centrally directed process of industrial reconstruction did emphasise the development of civilian rather than military enterprises, but in the heavy rather than the light industrial sphere.[16] As Table 5.1 shows, the share of war output in Soviet industrial production fell from 63.9 per cent in 1942 to 40.1 in 1945 but the main beneficiary of this was civilian heavy industry (the share of which rose from 20.5 to 34.5 per cent over the period) rather than light industry (the share of which rose from 15.6 to 25.4 per cent).

The main issues in sectoral policy in 1945 were therefore as follows. Should the wartime policy of concentrating reconstruction more on heavy than light industry be continued on defensive and long-run maximisation grounds? In other words, should the criteria of the civil war and of the pre-war Five-Year Plan periods be reinvoked in the reconstruction process? Alternatively would the heavy industrial might built up by the USSR before the war and substantially restored by 1945–6[17] allow a greater concentration of resources on the output of mass consumption goods in the post-war era? With the Soviet Union's increasing security from foreign attack was not the defence criterion less pressing? Would the Soviet population be prepared to endure another five years of privation to restore and further expand the USSR's heavy industrial base; would they tolerate consumption levels that might be even lower than those of the late 1930s? Such arguments formed a powerful case for giving a greater priority to the 'B' sector than previously, and they were voiced in the discussions leading to the formation of the Fourth Five-Year Plan.

Sectoral Policy Discussions before the Fourth Five-Year Plan

Sectoral policy like its regional counterpart was the subject of active debate in the period leading up to the announcement of the new Five-Year Plan in March 1946. The participants in the public debate included specialist officials and advisers of the command economy as well as the political leadership. The fact that such figures could challenge the primacy of heavy industry, if they took care to express themselves in the approved manner, is surely adequate testimony to the breadth of views that influenced policies in the command economy, at least in the 1940s. What was at one time, under NEP, official policy and later became 'rightist opportunism' was once again perfectly respectable!

The debate amongst the specialists was very much concerned with the defence and consumer satisfaction issues. In listing the tasks of post-war economic reconstruction the economist Kuzminov put the development of heavy industry (to provide a base for defence of the USSR) as the first priority. Raising living standards was only a secondary aim.[18] To Zelenovskii of Gosplan a prime task of planners in the post-war era was planning for the development of non-war industries, especially consumer goods and foodstuffs. He quoted Stalin's election speech of 9 February 1946, on the need to end rationing and raise living standards.[19]

Another Gosplan official, Korobov, took the opposite point of view and quoted the same speech of Stalin's as aid to *his* argument. He reiterated the defence argument but also raised the point that the primacy of heavy industry was one of the main features of the socialist path of economic development that distinguished it from the capitalist path.[20] The relative weight of heavy and light industry under various stages of socialism and/or communism became a common theme of discussion in later years. Korobov's contribution to the debate came in the form of a commentary on the new plan; for, as we shall see, the provisions of the plan in this sphere were not as clear as they might have been, perhaps a reflection of the disagreements that proceeded and accompanied it.

Debates amongst advisers influenced the Soviet political leadership and were reflected in public differences amongst members of the Politburo. Some of the ways in which 'officials' could hope to influence 'leaders' have been documented in the memoirs of the Finance Minister of the time, Zverev. His ministry had pressed the Council of Ministers to divert more resources towards light industry after the war, partly, no doubt, in order to stabilise the rouble.[21] The planners, however, did not at first pay 'over much attention to these suggestions' (*predlozheniya*), but the 'Central Committee' (probably, in reality, the Politburo and/or the Secretariat) intervened on the ministry's behalf. Zverev used the fact that he had allies within the leadership to put pressure on the branch ministries and Gosplan, who were clearly far less well disposed towards the claims of the 'B' sector than Zverev himself.

Amongst the Politburo membership it was Molotov and Zhdanov who publicly emphasised the need to expand consumer goods production in 1945–6. (Malenkov, whom many Western observers saw as an emerging opponent of Zhdanov,[22] showed no public evidence of his later post-Stalin advocacy of consumer goods.) In his speech on the twenty-eighth anniversary of the revolution in November 1945 Molotov emphasised the need to raise production of consumer goods and agricultural produce, with provision for defence needs coming only third in his list of priorities for the new plan.[23] One reason he advanced for paying greater attention to light industry was that 'Of course knowledge of the ways of life of other peoples has its benefits for our people and broadens their outlooks'.[24] The glimpses of life abroad offered to Soviet servicemen and officials during the war were an argument for expecting greater demands for consumer goods in the post-war era.

In its editorial commentary on Molotov's speech, the party's journal *Bol'shevik* reversed his priorities, calling 'above all' for the further development of heavy industry. The main argument for this was the need to rebuild Soviet defensive might and so secure the recent victory over fascism. Rising consumer expectations were firmly relegated to second place.[25] It may be assumed that other high officials of the party did not share Molotov's views on sectoral policy.

Bol'shevik was not alone in stressing the primacy of heavy industry. A *Pravda* editorial of 6 February 1946 made much the same point. However, it was only three days after that that Stalin made his last contribution to the debates before 1952. In his election speech he set forth grandiose targets for three heavy industrial sectors (steel, coal and oil) over the following fifteen years 'if not more', and more generally called for a trebling of industrial output over that period.[26] This speech did not, however, mark a decisive intervention in the heavy versus light industry debate. Its targets were too vague and far distant to signify that. Indeed both sides managed to quote Stalin's speech to 'prove' their points of view.[27] Certainly Zhdanov found no difficulty in reconciling Stalin's words with stressing the fact that under the new plan 'The task of developing and expanding the production of mass consumption goods is the subject of the special care and attention of the Soviet state'.[28]

Some Western observers, indeed, see a clear line-up of factions within the Politburo on the sectoral issue in 1946–7. McCagg, for example, puts the entire 'party revivalist-economic voluntarist' faction on the consumer goods side of the argument.[29] The quotations from Molotov and Zhdanov cited here certainly put them in that faction and Mikoyan's overall responsibility for trade matters made him an obvious consumer goods advocate. Yet, as we shall see later, Voznesenskii's position was, at best, uncertain. At worst he found himself far more in sympathy with 'steel eaters' like Kaganovich than with the 'consumerists'.

Sectoral Policy Decisions 1946–7

Whatever the line-up in the leadership, the new plan, as in the case of regional policy, represented what was essentially a compromise between the two points of view. The Fourth Five-Year Plan listed fifteen main tasks for the next five years.[30] The first and most important of these was the development of the heavy industrial base. This was justified in terms of maximising long-run returns ('the reconstruction and development of the entire economy of the USSR'). The need to further strengthen the Soviet Union's defensive capability was listed only fifth in the order of priorities. Second place was allocated to the development of agriculture and consumer goods output in view of the need to satisfy the expectations of the Soviet consumer ('to create an abundance of basic consumption goods'). In accordance with this statement of overall priorities the plan's detailed provisions represented a compromise between the advocates of heavy and of light industry as the basis for future growth. Light industry was to regain some but not all the ground it had lost to heavy industry during the war. The published plan document contained no overall targets for the 'A' and 'B' sectors, but Voznesenskii, in presenting it to the Supreme Soviet, announced a target of 17 per cent growth per annum for light industrial output.[31] That implied a target of 119 per cent growth in 'B' sector output over the five-year period and only 43 per cent in 'A' sector output. Given the decline in light industry output over the war years, this in turn indicated that the planned share of the 'B' sector in industrial output in 1950 was 34 per cent; in other words, well above its 1945 share of 25.1 per cent, but still less than its pre-war share of 38.8 per cent.[32] Table 5.2 summarises the plan and its fulfilment in the two main industrial sectors. Wartime proportions were not to be continued, but pre-war proportions were not to be *fully* restored during this plan period.

The relatively low growth target set for heavy industry is partly explained by the need to anticipate the costs of reconverting the heavy industry sector back to civilian production. As Table 5.3 shows, the output of most of the major *non-war* branches of heavy industry was to expand twofold or threefold over the five-year period.[34]

If we take as the base year for the plan targets 1946 or 1947 (i.e. after reconversion was substantially completed), it is clear that the 'A' and 'B' sectors were to grow at roughly the same rate. (See Table 5.4.) In so far as the

TABLE 5.2. Distribution of output by sectors of Soviet industry 1945–50

	Sector	
	per cent of gross industrial output	
Year	*'A'*	*'B'*
1945	74.9	25.1
1946	65.9	34.1
1947	66.0	34.0
1948	67.4	32.4
1949	70.6	29.4
1950 (Plan)	66.0	34.0
1950 (Actual)	72.4	27.6

Sources: See Note 33 on p. 163.

planners correctly anticipated the decline in heavy industrial output that resulted from the reconversion process, they were therefore putting forward targets that were more favourable to light industry than were the outcomes of the first three Five-Year Plan periods (all of which saw much faster growth of heavy than light industry). The rapid expansion planned in the output of certain consumer durables (see Table 5.3) is particularly noteworthy.[36] In so far as the planners underestimated the costs of reconversion (surely more likely than overestimating), they foresaw a faster expansion in 'B' sector than in 'A' sector output. In the light of these figures the claim of two economists in 1948 that the Five-Year Plan allowed for a much more rapid growth in heavy than in light industry must be viewed with scepticism. It will be argued that they were making a partisan point in the sectoral debate rather than giving information about the details of the plan.[37]

As to the sources of expansion for each sector, the published plan did not reveal how far new investment and labour recruitment were to be directed towards one sector rather than the other. Of the 153 billion roubles to be spent on construction and equipment work, 55 billion was to be used up by trusts of the ministries in charge of construction of enterprises in the fuel, war and other heavy industries. There was no equivalent ministry for light industry, which in any case was less capital-intensive. In the labour market the plan did make provision for extra pay to workers in some high priority branches of heavy industry (coal, metals and oil). It would not be unreasonable to suppose that the planned expansion of the 'B' sector would rely more heavily on small-scale local industry,[38] self-financed investment and non-centrally directed labour than would heavy industry. Even so the

TABLE 5.3. Output targets for major industrial sectors in the Fourth Five-Year Plan

Sector	*Planned output in 1950*	
	at 1940 = 100	*at 1945 = 100*
ALL INDUSTRY	148	161
GROUP 'A'	160	143
(i) *Fuel and Power*		
Electric power	170	190
Coal	151	167
Oil	114	182
Natural gas	261	256
(ii) *Ferrous Metals*		
Steel	139	207
Pig iron	131	222
Sheet metal[a]	136	209
(iii) *Machine-Building and Metal Processing*		
Machine-building	200	
(a) Transport Machinery		
Steam locos	241	27500
Diesel locos	6000	–
Electric locos	2444	–
Goods wagons	472	18250
Passenger coaches	247	52000
Lorries	315	624
Cars	1190	1313
Buses	163	575
(b) Other industrial machinery		
Metal equipment by weight	434	
Steam turbines[b]	299	
Hydro turbines: large	186	
medium	492	
small		
Electric motors: up to 100 kwt	239	562
over 100 kwt	290	281
Metal-cutting tools	127	193
Weaving machines	1388	125000
(c) Agricultural machinery		
Tractors	354	1454
Ploughs[c]	286	1294
Cultivators (harrows)[c]	255	9144
Sowers[c]	389	5206
(iv) *Chemicals*	150	144
Caustic soda	223	331
95% Soda ash	157	357
Mineral fertilisers	159	464

TABLE 5.3. *(continued)*

Sector	*Planned output in 1950*	
	at 1940 = 100	*at 1945 = 100*
(v) *Construction Materials and Forestry*		
Lumber	114	166
Saw-timber	112	265
Cement	185	569
Slates	199	488
Window glass	179	343
GROUP 'B'	129	219
(vi) *Textiles*		
Cotton fabrics	119/121[d]	290/296[d]
Woollens	133/14[d]	297/314[d]
(vii) *Other Pure Light Industry*		
Leather footwear	114	380
Rubber footwear	127	
Hosiery	120/130[d]	637/692[d]
(viii) *Foodstuffs*		
Meat	87	196
Animal fat	122	235
Vegetable oil	110	301
Sugar	86	462
Soap	124	380
(ix) *Consumer Durables*		
Clocks	264	2466
Radios	578	6655
Bicycles	412	
Cameras	149	
Domestic sewing machines	257	
Motorcycles	2014	2872

Sources: See Note 35 on p. 163.

[a] '*Prokat iz chernykh metallov*'.
[b] By capacity.
[c] Tractor-drawn.
[d] Figures refer to the revised targets of the decree of 23 December 1946.

All-Union Ministries of the Light and Textile industries were to receive 5.99 billion and 2.18 billion roubles respectively in investment funds over the five-year period; that is, 5.2 per cent of all industrial investment in the USSR, compared to 2.4 per cent during the war and 7.9 per cent during the Third Five-Year Plan period for the entire pure light industry sector (*not* just these two ministries).[39] The foodstuffs ministry was to be allocated 5.6 billion roubles of investment funds over the plan period.[40]

TABLE 5.4. Planned increases in output of sectors after completion of 'basic' reconversion to civilian needs (1946 or 1947)

	Fourth Five-Year Plan Target for 1950	
	at 1946 = 100	*at 1947 = 100*
All industry	192	159
'A' Sector	195	158
'B' Sector	193	158

Sources: As Table 5.2.

These investment figures seem to correspond to the policy line enunciated in the preamble to the plan and in the sectoral output targets. That policy was a compromise between the aim of the maximising industrial output in the longer term by basing reconstruction on civilian heavy industry and that of satisfying the Soviet consumer's long-frustrated demand for consumption goods. The low priority accorded to the defence factor was reflected in the dramatic decline in the output of war industries in 1946.[41]

Sectoral policy decisions in the 1940s did not, however, end with the Five-Year Plan. The following two years saw, first, a re-emphasising of the priority of light industry and then a partial reversal of this trend with the decision to boost arms production from late 1947.

It became clear during 1946 that light industry was not performing satisfactorily. The foodstuffs industry's problems were, of course, mainly due to the drought of 1946 and could thus reasonably be attributed to poor agricultural performance. The rest of the 'B' sector was, however, having its own problems, often connected with the process of reconversion. For example, a combined plenum of the Moscow City and Province Party Committees in March 1946 discussed such difficulties in the local textile industry.[42] The local textile mills produced about 40 per cent of the total Soviet output, but were finding it difficult to make the conversion from producing uniforms for the armed services to producing clothing for the civilian market in the required quantity and quality. The importance attached to these problems was illustrated by the attendance at the plenum of Kosygin (then the Council of Ministers' 'overlord' for the light industrial sector). *Pravda* followed this up some three months later with an editorial stressing the need to expand the output of textiles to satisfy consumer demand.[43]

It was in response to these problems that the leadership sought to strengthen the light industrial sector by means of two decrees published in

November and December 1946.[44] The aim of these measures was to pump more resources into the 'B' sector, and also to raise the output plan targets for those industries. In other words, they amended the Five-Year Plan's targets in favour of light industry and in favour of the consumer satisfaction criterion.

The Council of Ministers decree of 9 November 1946 on co-operative trade and production organisations judged the work of these bodies 'extremely unsatisfactory' and called for increases in production of mass consumption goods by co-operative and local industry. The decree of 23 December deemed the work of the All-Union ministries of the textile and 'pure' light industry to increase output 'unsatisfactory' and insufficient to meet the targets set out in the Five-Year Plan.

The remedies announced for both small and large-scale industry were, first, to transfer resources from other sectors to the 'B' sector; second, to improve the work of the relevant controlling organisations; and, lastly, to raise the output targets set for those organisations in 1947 (and throughout the Five-Year Plan period for the All-Union ministries).

The decree on co-operative industry called for these small enterprises to be freed from orders for goods from other industrial organisations and to 'transfer on a large scale to the production of mass consumption goods for the population'. The decree on All-Union industry not only provided for a 25 per cent greater allocation of investment funds to the light and textile industry ministries, a special recruitment of 300,000 workers in 1947, more funds for housing construction for them, and increased supplies of machinery and buildings, compared to the original provisions of the Five-Year Plan—it also positively ordered other sectors to give up resources to these two ministries. For example, investments in other branches of industry (specified in an unpublished appendix) were to be cut back; ministries that had recruited workers from pure light and textile plants in the past were now to return them; various ministries and local Soviets were to return buildings to the light and textile ministries on this basis. These transfers could only reflect a greater emphasis on light as against heavy industry.

This change in emphasis was reflected in the demands made in these decrees for other ministries and organisations to improve supplies to light industry. Both decrees accused local organisations of treating light industrial enterprises as of secondary importance and *Pravda* in its editorial commentary on the December decree stressed that increasing consumer goods output was now 'one of the highest priority and most urgent tasks of the whole state'.[45] Local Soviets and party organisations were thereby instructed to give greater priority to light industrial enterprises than previously. The light and textile industry ministries themselves were also enjoined to improve their work and were obliged to report monthly direct to the USSR Council of Ministers on the progress of construction and reconstruction work. As we have seen, in wartime such a provision had been made only in the highest priority sectors.

Finally, these two decrees amended the targets for light industrial output specified in the Five-Year Plan. Local and co-operative industry in 1947 was to produce nearly 30 per cent of the total Soviet output of shoes and nearly 20 per cent of hosiery output (although they were to produce less than 3 per cent of fabrics production). The All-Union targets for fabrics and hosiery output in 1950 were raised by between 2 and 9 per cent. These decrees did not simply seek to enable light industry to fulfil its existing obligations: they made more concessions to consumer demand than had been made in March 1946.

The aim of these measures was to amend the sectoral policy of the Five-Year Plan. Their detailed provisions clearly show that neither the original plans nor these amendments were window-dressing designed to fool foreign observers and Soviet consumers. If that were the aim of these plan targets, then why go to the trouble of not only setting new ones but devising what seemed to be realistic means of achieving them? As we shall argue in the next chapter, these targets for light industry were, by and large, not met *not* because they were never intended to be, but *in spite of* the leadership's clear intentions that they should be fulfilled.

It is, however, commonly assumed that this policy was reversed in about December 1947.[46] It was in this month that, faced with a worsening international situation, the USSR apparently decided to rearm. Such a decision would have presumably entailed an upward revision of targets for both the military and the civilian sectors of heavy industry. Yet, as is the case with a number of assumptions about the cold war era, there is no hard evidence that such decisions were made. Malenkov told his Cominform colleagues that they could and would be made, but not all apparently agreed with his views.[47]

Even Soviet economic historians have, however, admitted some decision to rearm around this date.[48] The question is really one of the impact of this decision on the A/B balance. The arms build-up, according to official budget figures, was meek in comparison to that before the war (which many have considered inadequate). Only 20 per cent of the state budget for 1950 was allocated to defence compared to 32.6 per cent in 1940.[49] The only intersectoral transfers of investment funds known to the author were some 450 million roubles from pure light to heavy industry in 1948, balanced by a transfer of 343 million roubles in the opposite direction in the following year. These figures, in any case, represent less than 5 per cent of investments in light industry over the Five-Year Plan period. That there was in fact relatively little change of emphasis planned in the direction of civilian industrial expansion is perhaps illustrated by the lack of any 'explosion' in 'A' sector growth after 1947. As we shall argue in the next chapter, even if there was any large-scale change of priorities in favour of heavy industry, it had remarkably little effect on its actual growth rate. Leadership decisions on sectoral policy, the evidence suggests, remained substantially more favourable to the 'B' sector in terms of growth targets than either before or during the war.

Sectoral Policy Discussion 1946–52

As in the case of regional policy, the taking of decisions by the leadership did not preclude further debate on sectoral matters. The debates over sectoral policy over the period 1946–52 may be divided into those conducted in more purely economic terms and those conducted in terms of political economy and invoking more general theoretical or 'ideological' arguments (notably the debate on proportions at various stages during the transition from socialism to communism). In addition debates over other issues, such as investment criteria and pricing policy, were relevant to the sectoral issue.

The economic criteria used to argue for priority for heavy industry were mostly those of defence and long-run maximisation of economic growth. In spite of the relaxation of international tension immediately after the war and the greatly enhanced defensive might of the USSR in 1945, officials and academics were still stressing the need for the continued development of heavy industry to boost Soviet military capabilities. For example, as we have seen, the Gosplan official Korobov, in commenting on the new plan in 1946, quoted Stalin on the need to develop heavy industry for defence. Many of the academic books and articles on the war economy published at this time echoed this theme. In a book published in December 1946, Gatovskii described heavy industry as 'the basis of the technical and economic independence of our nation and of its military might'.[50] A number of other academic commentators at this time stressed not only the defensive aspect, but also the need to reconstruct civilian heavy industry as the base of post-war industrial growth. Arakelyan described the expansion of heavy industry as a lever for the future development of other sectors of the economy.[51] Lokshin, in an article uncompromisingly titled 'The Soviet Method of Industrialisation' asserted that 'In the circumstances of the post-war economy heavy industry will preserve its leading role',[52] a statement that was not fully borne out by the plan directives. The 'leading role of heavy industry' was also stressed by Braginskii and Vikentev in 1948. As has already been noted, they even claimed that the Fourth Five-Year Plan envisaged faster growth in the 'A' sector than in the 'B' sector. That flatly contradicted all the official pronouncements of 1946! The senior editor of Gosplan's *Planovoe Khozyaistvo* justified priority for heavy industry after the war on the ground that its leading branches had suffered most of all from wartime destruction,[53] ignoring the fact that the 'B' sector had suffered far more from simple neglect rather than destruction during the war. The next issue of the journal carried an editorial listing the main tasks of post-war planning as the development of heavy industry and railway transport, with no mention of consumer goods.[54] The senior editor, G. P. Kosyachenko, was clearly an influential figure in Gosplan; indeed he came to head the state planning organisation for a time in 1953. He clearly tended to favour heavy industry in the sectoral debate. The major part of the heavy industrial lobby consisted of academics and Gosplan officials, although some of the former took an opposing position.

It was only in 1947 that *Planovoe Khozyaistvo* allotted much space to the light industry point of view and that article was little more than a belated reinforcement of the December 1946 decree.[55] It was in the columns of *Pravda* that the need to produce more and better consumer goods was more frequently emphasised.[56] That emphasis may have reflected a consumer goods orientation on the part of highly placed officials in the light industry and trade sector. For example, Mikoyan went so far as to quote Stalin at the 1952 Party Congress to the effect that 'socialism can only triumph on the basis of . . . an abundance of . . . all kinds of consumer goods'.[57]

It was to this question of the sectoral balance of Soviet industrial growth during the 'transition from socialism to communism' that many academic economists addressed themselves in the late 1940s and early 1950s. This debate is sometimes called after A. I. Notkin, who argued that radical revision of the old relationship between heavy and light industry (favouring the former) was not necessary to the transition to communism that people were being encouraged to regard as immanent.[58] However, this theme had been touched on several years earlier. It was commonly agreed that the process of industrialisation that marked the transition from capitalism to socialism was based on priority to heavy industry. This was the 'Bolshevik method of industrialisation' and to argue otherwise was 'Trotskyite' and/or 'Bukharinite'.[59] It was quite another step, however, to argue as Petrov did in 1947, that the theory of equal growth of all sectors was 'counter-revolutionary', irrespective of the stage of social development to which it was applied. To Petrov faster growth of the 'A' sector was as characteristic of socialist society and economy (the stage at which the USSR was supposed to be in 1947) as of a society in transition from capitalism to socialism.[60]

In January 1948 Ostrovityanov instructed the newly reorganised Institute of Economics of the Academy of Sciences to devote its energies to the working out of the 'laws' governing the development of socialist economies, paying special attention to the issue of sectoral balance at different stages in their development and movement towards communism.[61] This became associated with the writings of a new textbook on political economy that was to be an authoritative doctrinal statement on this and other problems. The debates on this new textbook were therefore crucial to the education of the next generation of Soviet economists, planners and administrators, as well as directly relevant to the formation of sectoral policy at the highest levels.

The discussions on the new textbook formed the central theme of a meeting of the party organisation of the Institute of Economics in 1950.[62] Notkin argued that, although some branches of light industry would have to be expanded rapidly to create the material basis for communism in the USSR, the overall expansion of the 'B' sector would continue to be slower than that of the 'A' sector. The author of the report of the discussion, Anchishkin, on the other hand, emphasised that an abundance of consumer goods was essential to the building of communism. He admitted, however, that he had been attacked by a number of other speakers and conceded that an abundance of such goods did not require the same overall level of output under socialism or communism as under capitalism, due to the better distributive system operating in socialist society. V. F. Vasyutin attacked even this concession to the heavy industry point of view, claiming that the conference had spent far too much of its time discussing distribution rather than production. If the level of output were high enough for the establishment of communism, the distribution problem would be easily solved. The meeting showed that some economists saw light industry as a greater priority in the next phase of economic policy than did others.

A plenary session of those concerned with the new textbook was convened in November 1951. The points raised in these discussions formed the basis for Stalin's 'Economic Problems of Socialism in the USSR'. This was published in *Bol'shevik* and *Pravda* in October 1952 and subsequently as a pamphlet in many millions of copies. Much of the pamphlet was not particularly enlightening, although it did provide a number of speakers at the Party Congress with suitable quotations to back up various points of view! The main relevance of this theoretical work to the sectoral issue was in an appendix containing Stalin's reply to L. D. Yaroshenko. Yaroshenko had written to the Politburo asking why his views had not been included in Stalin's summary of the November 1951 discussions. Stalin's published reply was that Yaroshenko had made a number of serious errors in his contribution to the debate, that he held an un-Marxist point of view and that he was being presumptious in his request to write the textbook himself.[63] Yaroshenko's major error, according to Stalin, was to assert that all that was necessary to make the transition from socialism to communism was to ensure the 'rational organisation of productive forces'. Stalin reiterated the point that the transition was a matter of production, not merely organisation. The building of communism required

> the uninterrupted growth of all social production with the growth of the production of means of production predominating. The predominating growth of the production of means of production is necessary not only because it would provide equipment for its own enterprises and also for enterprises of all other branches of the economy, but also because, without it, it is altogether impossible to realise expanded reproduction.[64]

In other words Stalin was putting the long-run maximisation case for 'A' sector dominance. His contribution closed the debate until his death, but its long-term influence may be judged by the fact that within a year Malenkov was advancing the sort of consumer goods view that Stalin was implicitly attacking.

It has been suggested that Stalin's 'Economic Problems' was meant as a point-by-point refutation of Voznesenskii's book on the war economy published in 1947. Some have even speculated that 'Comrade Yaroshenko' was a surrogate for Voznesenskii, and his proposed textbook Voznesenskii's unpublished 'Political Economy of Communism'.[65] The chairman of Gosplan had certainly favoured the use of economic levers like the price mechanism for determining economic proportions. If one accepted Voznesenskii's belief that the 'law of value' operated under socialism, albeit in a 'transformed' way (to use the contemporary jargon), then, Stalin argued, how could heavy industrial growth be justified, as concentration on light industry was more 'profitable'?[66] Stalin's answer was that the law of value did not so operate, in spite of the fact that it had been Stalin himself who some years before had asserted that it *did*!

Voznesenskii has been closely associated with the promulgation in 1948 of wholesale price rises averaging 60 per cent for heavy industrial goods. In so far as these were designed to eliminate budgetary subsidies to heavy industry, they seem to confirm Voznesenskii's position as a consumer goods advocate. If one assumes this, then the partial reversal of the price increases in 1949–50 can be attributed to his fall. It could also be reflected in the dismissal of Fedoseev and two colleagues from the editorial board of *Bol'shevik* in July 1949 for favouring Voznesenskii's views.[67]

Voznesenskii's 1948 measures (if indeed they were primarily his responsibility) could, however, also be interpreted as *benefiting* heavy industry. The aim of the increases was initially to enable 'A' sector enterprises to make profits and thereafter to force them to become more efficient, assessing that efficiency in terms of their ability to meet targets for profitability, not just physical output.

In so far as price reforms made heavy industry more efficient, they surely benefited the 'A' sector! A Soviet economist has explained the partial reversal of increases in 1949–50 as a reflection of the new-found profitability of some heavy industrial branches. Only by cutting prices again could industries making profits of 20 per cent or more be induced to seek other ways of becoming more efficient and cost-conscious.[68] It may be that the fall of Voznesenskii and some of his allies owed more to his popularity amongst economists and politicians and to his ties with the Zhdanov faction within the leadership (and thus the enmity of Malenkov, Beria, Khrushchev and Suslov[69]) than to his views on economic policy. Indeed Stalin, in replying to a letter from Notkin in 1952, was at pains to attack those who denied the utility of the notion of profitability in a socialist economy. Profitability of the individual enterprise and branch, he argued, 'must be taken into consideration both in planning construction and in planning production. This is the ABC of our economic activity at the present stage of development'.[70] Stalin may have been suspicious of Voznesenskii's ambitions and popularity but there is little evidence that the two men disagreed fundamentally over sectoral or pricing policy. In accusing 'Comrade Yaroshenko' of favouring the consumer goods sector, Stalin was, perhaps, using a convenient stick with which to beat Voznesenskii and his surviving sympathisers. As we have argued in this chapter, considerations of the 'law of value' (that is, of profitability) do not rule out investment in heavy industry; it depends on the planner's time horizon. Whatever Stalin branded him as, it is, at the very least, unproven that Voznesenskii was a defender of the 'B' sector interest.

The debate over investment criteria cited in Chapter 3 also had its implications for sectoral policy. If investment projects were to be assessed mainly in terms of initial cost and shorter-run private returns, then light industrial investment would be favoured. Longer-run profitability criteria, social returns and political considerations had to be invoked to justify investment in heavy industry even after the 1949 price increases. In his reply to Notkin, Stalin seemed to favour a combination of private profitability calculations with overriding political and social considerations. That seemed closer to the position of Mstislavskii than to the marginalism of Novozhilov, but without totally rejecting the latter's point of view. Stalin's views were in fact very close to those advanced by Khachaturov in summarising this debate in 1950. The results of the investment criteria debate were thus rather inconclusive for sectoral policy. The discussions did, however, suggest that light industry was not without its advocates amongst the 'profit criterion' faction, and heavy industry not without its supporters amongst the 'political criteria' faction.

Sectoral Policy Decisions 1951 and After

In the interregnum between the ending of the Fourth Five-Year Plan period and the announcement of the new plan the impact of the light industry lobby could still be felt in leadership decisions. In 1951, for example, increased targets for consumer goods output in the annual plan were announced. That they were supposedly 'on the initiative' of Comrade Stalin[71] was presumably to give them greater weight and legitimacy.

However, with the publishing of the new plan in October 1952, the pro-heavy industry stance of Stalin's 'Economic Problems of Socialism' (which had widespread support amongst officials and advisers) was once again enshrined as official policy. The 'A' sector was to expand at 13 per cent per annum over the Fifth Five-Year Plan period and the 'B' sector at 11 per cent. Even so, that represented a much greater concession to light industry than the growth proportions of the 1930s.

In any case, within a year of the announcement of this new plan, its targets were substantially revised in favour of the consumer goods industries. This move, announced in August 1953, was very much associated with Malenkov and was heralded by various reforms that began only a month after Stalin's death.[72] The debates over the sectoral balance continued for the next few years in the context of Malenkov's rivalry with Khrushchev. Yet, as we have shown, these debates began long before the death of Stalin, although Malenkov's position on the issue does seem to change. In the 1940s his political opposition to 'consumerists' like Zhdanov and Molotov and his base of support amongst industrial managers led him closer to the 'steel eater' point of view. Indeed, as we shall see, his later advocacy of consumerism does seem rather politically inept, as all the evidence points towards a pronounced heavy industry bias amongst the majority of ministerial controllers of industry. The advocates of light industry clearly had some success in influencing sectoral policy decisions but, as in the case of regional policy, they were to find it far more difficult to secure the implementation of those decisions. That was very much the province of the line officials of the command economy, where Malenkov enjoyed much support.

6 The Implementation of Sectoral Policy after the War

The Fulfilment of the Fourth Five-Year Plan

The evidence presented in Chapter 4 points to the dominance of production branch over regional interests in the implementation of economic policy. It also shows how the leadership's policies were distorted in the process of their execution. It remains to be asked, however, whether these distortions were due to the absence of an administrative agency with an interest in implementing regional policy and the direct control over production to enable it to do so. The interests of heavy and light industry were so 'represented' by branch ministries. By looking at sectoral policy we should therefore be in a better position to decide whether the distortions in regional policy were due to the absence of ministerial defenders or to a more basic feature of the Soviet political system—that the implementors of economic policy had more influence over its direction than did the nominal policy-makers, the leadership.

To determine this we must examine the extent to which the leadership's decisions of 1946 (and 1947) were carried out. The distribution of gross industrial output amongst the major sectors of the Soviet industry in 1950 showed a much greater bias towards heavy industry than had been laid down in the Five-Year Plan. The output of 'B' sector industry more than doubled over the plan period, but only managed to surpass its pre-war levels in 1949.

As Table 6.1 shows, the target for this sector was not met; only 95 per cent of the planned output level for the 'B' sector in 1950 was achieved. In contrast, the plan for industrial output as a whole was overfulfilled by 17 per cent and that for heavy industry by 28 per cent. In the fulfilment of the plan its original aims on sectoral policy were not achieved. Instead of light industry accounting for 34 per cent of industrial output in 1950 (as planned), it accounted for only 27.6 per cent. This was very little more

TABLE 6.1. Fulfilment of the Fourth Five-Year Plan's output targets for industrial sectors

Sector	*1* *Output (at 1940 = 100) in 1950*	*2*	*3* *Output (at 1945 = 100) in 1950*	*4*	*5* *per cent fulfilment of plan*
	Plan	*Actual*	*Plan*	*Actual*	
ALL INDUSTRY	148	173	161	189	117
GROUP 'A'	160	205	143	183	128
(i) *Fuel and Power*					
Electric power	170	189	190	211	111
Coal	151	157	167	175	104
Oil	114	122	182	195	107
Natural gas	261	179	156	176	69
Manufactured gas					
(ii) *Ferrous Metals*					
Steel	139	149	207	223	108
Pig iron	131	129	222	218	98
Sheet metal[a]	136	160	209	246	117
(iii) *Machine Building and Metal Processing*					
Machine building	200	234			117
(a) Transport machinery					
Steam locos	241	108	27500	12300	45
Diesel locos	6000	2500			42
Electric locos	2444	1130			46
Goods wagons	472	164	18250	6350	35
Passenger coaches	247	87	52000	18240	35
Lorries	315	217	624	429	69
Cars	1190	1171	1313	1292	98
Buses	163	100	575	354	62
(b) Other Industrial Machinery					
Metal equipment (by weight)	434	469			108
Steam turbines	299	245			82
Hydro-turbines;					
— large	186	80			43
medium / small	492	152			31
Electric motors:					
up to 100 kwt	239	302	562	709	126
over 100 kwt	290	510	281	494	176
Metal-cutting tools	127	121	193	184	95

TABLE 6.1. *(continued)*

Sector	*(1)* *Output (at 1940 = 100) in 1950* *Plan*	*(2)* *Actual*	*(3)* *Output (at 1945 = 100) in 1950* *Plan*	*(4)* *Actual*	*(5)* *per cent fulfilment of plan*
Spinning machines					
Weaving machines	1388	483	125000	43500	35
(c) Agricultural					
Tractors	354	369	1454	1516	104
Ploughs	286	317	1294	1434	111
Cultivators tractor (harrows) drawn	255	306	9144	10988	120
Sowers	389	553	5206	7356	141
(iv) *Chemicals*	150	196	144		131
Caustic soda	223	171	331	253	77
95% Soda ash	157	140	357	317	89
Mineral fertilisers	159	172	464	500	108
Synthetic dyes					
(v) *Construction Materials and Forestry*					
Lumber	114	108	166	158	95
Saw-timber	112	142	265	337	127
Cement	185	180	569	553	97
Slates	199	265	488	650	133
Widow-glass	179	172	343	330	96
GROUP 'B'	129	123	219	208	95
(vi) *Textiles*					
Cotton fabrics	119/121[b]	99	290/296[b]	241	83/81[b]
Woollens	133/140[b]	130	297/314[b]	290	97/92[b]
(vii) *Other Pure Light Industry*					
Leather footwear	114	96	380	322	85
Rubber footwear	127	158			125
Hosiery	120/130[b]	98	637/692[b]	520	82/75[b]
(*vi & vii Pure Light Industry*)					
(viii) *Foodstuffs*					
Meat	87	104	196	235	120
Animal fat	122	149	235	287	122
Vegetable oil	110	103	301	280	93
Sugar	86	115	462	621	134
Soap	124	117	380	356	94

TABLE 6.1. *(continued)*

Sector	(1) Output (at 1940 = 100) in 1950 Plan	(2) Actual	(3) Output (at 1945 = 100) in 1950 Plan	(4) Actual	(5) per cent fulfilment of plan
(ix) *Consumer Durables*					
Clocks	264	271	2466	2300	103
Radios	578	670	6655	7712	116
Bicycles	412	254			62
Cameras	149	73			49
Domestic sewing machines	257	287		112	
Motorcycles	2014	1836	2872	2617	91

Sources: See Note 1 on p. 164.
[a] *Prokat iz chernykh metallov* (if the plan uses a wider definition, overfulfilment is greater).
[b] Figures refer to the revised targets of 23 December 1946.

than its abnormally low 1945 proportion of 25.1 per cent and well short of its none too high 1940 share of 38.8 per cent. Soviet economic historians have admitted that the light industrial sector was the poor cousin of the Fourth Five-Year Plan period and failed to meet the demand made on it by Soviet consumers.[2]

The bias in plan fulfilment in favour of the 'A' sector indicated by the aggregate data is also reflected in the data for more specialised industrial branches presented in Column 5 of Table 6.1. It is true that some branches of heavy industry did *not* fulfil their output plans for 1950, notably natural gas, several branches of the machine-building industry, and some construction materials branches. But the major branches—such as coal, oil, electric power, metals and machine building as a whole—overfulfilled their targets. The machine-construction sector alone had accounted for more than a quarter of Soviet industrial output in 1938;[3] it overfulfilled its target in the first post-war plan by 17 per cent, and formed the basis for the overfulfilment of heavy industry's tasks.

The claim that industry as a whole produced 17 per cent more than was planned must be viewed with scepticism. Very few of the individual branches of industry specified in the plan document fulfilled their plans by more than 117 per cent; many more fulfilled them far more modestly or failed to fulfil them at all.[4] The contrast between the overfulfilment of plans for most heavy industrial branches and the underfulfilment of those for light industrial branches, however, is clear from this data. Of the main textile industries none fulfilled its plan and yet these had produced more than a quarter of light industrial output in 1938. Amongst the other non-food light industries only the targets for rubber footwear and a few consumer durables (such as clocks, radios and sewing machines) were met. If one considers the revised targets for consumer goods output announced in December 1946, the underfulfilment of plans in some branches becomes even more pronounced. Only in the sphere of foodstuffs production did plan fulfilment approach the satisfactory. The targets set for those industries were very modest in view of the low consumption levels of 1940 and (especially) 1945. The plan and the decree of December 1946 had shown originality in its stress on the need to expand the output of clothing and consumer durables. It was in these branches that the plan was most seriously underfulfilled.

Local and co-operative industry could not compensate for the (relative) failures of the All-Union ministries in the 'B' sector. Its total output in 1950 was 50 per cent above its 1940 level, but, as the official report on the fulfilment of the plan noted, such a growth rate was insufficient to meet the rising demand for consumer goods, especially in the assortment and quality of the goods produced.[5] Problems of quality and assortment also bedevilled the All-Union ministries in this field.[6]

It will be argued that this bias in plan fulfilment can be traced to differential supplies of various inputs to different sectors. However, as we have noted, some see it as no more than a reflection of a change in leadership policy, from about December 1947, in favour of civilian and military branches of heavy industry. If leadership policy was substantially altered from the position of the plan and the December 1946 decree, and if the leadership's commands were normally carried out, then one would expect a much greater bias in plan fulfilment in favour of the 'A' sector from 1948. Yet a comparison of annual growth rates of both sectors over the period 1945–50 shows no such pattern of sudden change. As Table 6.2 shows, after the reconversion problems of 1945–6 the annual growth rate of heavy industry was remarkably steady at between 23 and 29 per cent. That for light industry did drop substantially in 1949, but recovered again in 1950. An examination of the performance of major light industrial sectors, in relation to the revised targets of December 1946 (which were produced for each annual division of the five-year period) only reveals consistently poor performance in the cotton industry and a sudden decline in the woollens sector in 1950 (rather than 1948 or 1949). (See Table 6.3.) There is no sign in any of this data of the heavy industrial growth rate suddenly rising in 1948 or of plans for 'B' sector industries being especially underfulfilled in the same year. At the very least one can guess that actual

TABLE 6.2. Annual growth rates of sectors of Soviet industry 1945–50

	Output in per cent of previous year	
	'A' Sector	*'B' Sector*
1945	82	109
1946	73	114
1947	123	122
1948	129	121
1949	125	108
1950	126	115

Source: See note 7 on p. 164.

TABLE 6.3. Fulfilment of annual output targets for 'B' sector industries set in December 1946

	Percentage fulfilment of target	
	Cotton Fabrics	*Woollens*
1947	95	107
1948	92	112
1949	87	111
1950	81	92

Sources: See Note 8 on p. 164.

industrial proportions were the product of something else than leadership decisions, whatever their details and magnitude. The poor performance of the large cotton industry, for example, is perhaps better explained by the fact that, by as late as January 1949, it had regained use of only 35 per cent of its pre-war production capacity.[9]

The defence and consumer satisfaction factors may not, therefore, have been an important influence on the pattern of plan fulfilment; neither was the desire to maximise short-run returns on investments. Using the incremental capital-output ratio technique described in Chapter 4, it can be shown that the yield (in output terms) per unit of investment from light industry over the five-year period was three to four times greater than for heavy industry (see Table 6.4).

TABLE 6.4. Relative yields of funds invested during Fourth Five-Year Plan by sectors

	Relative ICOR (USSR = 100)
ALL INDUSTRY	100
'A' Sector	132
'B' Sector	37

Source: See Note 10 on p. 164.

A policy of maximising returns over the five-year period would therefore have dictated a much greater emphasis on light industry than was evident by 1950.

Sectoral Policy Fulfilment 1951 and 1952

As Table 6.5 shows the growth rates that light industry achieved in 1951 and 1952 were not far behind those achieved by the 'A' sector. They were, however, no greater than the 'B' sector had attained for most of the previous five years. Heavy industrial growth did slacken off in 1951 and 1952 as the overall industrial growth rate declined due to the end of reconstruction. The plan fulfilment figures in Table 6.5 refer to the targets of the Fifth Five-Year Plan, which were, however, only announced in October 1952. The marginally better fulfilment in the 'B' sector does not necessarily, therefore, reflect obedience to leadership commands. Indeed,

TABLE 6.5. Output of main industrial sectors 1951 and 1952

	Output in per cent of preceding year			
	'A'		*'B'*	
	Actual	*Fulfilment of plan*	*Actual*	*Fulfilment of plan*
1951	116.7	103.3	116.0	104.5
1952	112.2	99.3	110.7	99.7

Source: See Note 11 on p. 164.

the most likely explanation of it is the fact that, whilst heavy industry had been basically reconstructed by about 1948, many branches of light industry were still being reconstructed in the early 1950s. As we have argued elsewhere in this book, investment yields in reconstruction are potentially greater than those in new construction, as it is often cheaper to resurrect damaged or simply unexploited machinery and buildings than to build them from scratch. As we shall see, many of the problems of supplies of new resources to light industry continued well into the 1950s.

Sources of Overfulfilment and Underfulfilment

Following the methodology used in Chapter 4, we shall now seek to isolate the shortfalls and surpluses in the supply of funds and human and material sources to various branches and to assess the extent to which they might account for the underfulfilment of output plans for light industry and the overfulfilment of the tasks set for heavy industry.

Capital Allocations and their Utilisation

The distribution of investment funds amongst the major sectors of Soviet industry over the Fourth Five-Year Plan period is outlined in Table 6.6. Unfortunately the details of the planned distribution of these funds were not made public. However, it is known that the plan for expenditures on capital construction in industry as a whole was overfulfilled by 22 per cent, and that that for those on capital construction in pure light industry was underfulfilled. The orginal plan called for 5.2 per cent of industrial investment to be directed towards the enterprises of the Ministries of the Light and Textile Industries.[13] The decree of December 1946 raised this target to 6.5 per cent. In fact only 4.2 per cent of industrial investment went to the entire pure light industrial sector: that is *including* investment in local and co-operative enterprises as well as those under All-Union control. The investment plans in this sector must therefore have been underfulfilled by *at least* 1.5 per cent (taking the original target) or 21 per cent (taking the December 1946 target). The planned investments by the old Ministry of the Foodstuffs Industry amounted to 3.6 per cent of the target for all industry. In practice the 'new' ministry (which incorporated the former Ministry of the Confectionery Industry), along with the Ministries of the Meat and Dairy Foods Industry and of the Fishing Industry and local and co-operative enterprises, together acounted for 7.9 per cent of investment

TABLE 6.6. Distribution of industrial investment by sectors 1938–50

	Proportion of state and co-operative investment in industry during			
	3rd-Plan Period 1938–41	*War 1941–45*	*4th-Plan Period 1946–50*	
			Actual	*Planned*
GROUP 'A'	84.5	93.3	87.9	
Ferrous Metals	7.1	12.0	10.9	
Chemicals	4.7	3.6	3.7	
Oil and gas	7.4	7.9	11.5	
Coal	6.2	9.8	15.5	
Electric power	7.8	6.3	7.6	
Machine-building	33.5	34.4	16.4	
Construction and construction materials	4.6	5.5	9.4	
Forestry, paper and wood-processing	3.5	2.5	4.8	
GROUP 'B'	15.5	6.7	12.1	
'Pure' light industry	7.9	2.4	4.2	5.2[a]/6.5[b]
Foodstuffs	7.6	4.3	7.9	
ALL INDUSTRY	100	100	100	100

Sources: See Note 12 on p. 164.

[a] These figures are for investments by All-Union ministries only.
[b] These figures are as amended by 23 December 1946 decree of the CM.

in industry over the plan period. That presumably left very little room for any overfulfilment of plans for capital expenditure in the foodstuffs sector.

A comparison of the shares of the various sectors in Soviet industrial investment over time shows that the 'B' sector's share over the period 1946–50 was well above its wartime level, but had by no means regained its pre-war (Third Five-Year Plan) level. This is especially true for pure light industry, whose share, at 4.2 per cent, was well below its 1938–41 level of 6.7 per cent. The foodstuffs sector, however, did manage to regain its pre-war proportion of investment expenditures (see Table 6.6).[14] This fact must partly explain the better fulfilment of output plans in this sector than in the rest of the 'B' group industries.

If there is very little positive evidence that the leadership actually sought to divert investment funds away from light industry on a consistently large scale, why were the planned capital allocations for this sector not realised?

The volume of capital construction in light industry grew very rapidly after the war, achieving in 1946 a level of 67 per cent above that of 1945 (compared to a figure for all industry of 17 per cent). In reporting this fact, however, the editors of *Planovoe Khozyaistvo* still listed the pure light and textile industries as one of six priority sectors in the investment programme for 1947.[15] The rate of construction and reconstruction in these branches was still far from satisfactory. One reason for this was the failure of the light and textile industry ministries to make use of all the appropriations available to them for capital construction, at least in 1946–7.[16] The Minister of Finance, Zverev, noted that the textiles ministry had 23.5 million roubles of funds for capital construction 'illegally immobilised' during the first half of 1946.[17] As late as March 1949 deputies to the Supreme Soviet were criticising another ministry in this sector (that of the Meat and Dairy Industry) for not utilising all of the funds allocated to it for capital investment.[18]

One reason for this failure to use up the funds allocated was that there simply were not sufficient resources available on which to spend the funds. Zverev blamed the slow use of appropriations in 1946 on the inadequacies of the 'system of material-technical supply'. One reason for such problems in the 'B' sector was surely that there was no specialist construction ministry for light industry as there was for the fuel industry (until December 1948), for military and naval industries (later replaced by one for the machine-building industry), and for the rest of heavy industry. It is evident from the December 1946 decree cited above that construction work in the light industrial sector was primarily the responsibility of the branch ministries themselves, and also of the Ministry for the Construction of *Heavy* Industrial Enterprises. The construction of light industry enterprises came fairly low on the latter's list of priorities.

Much of the construction work in the 'B' sector was done by trusts of the light, textile, foodstuffs and allied ministries themselves who depended on other (heavy industrial, construction and transport) ministries for supplies to construction projects in a way that the specialist construction ministries did not.[19] The shortage of supplies must have been one reason for the 'B' sector ministries' failure to utilise the funds allotted to them.

The pricing structure for Soviet industrial output militated against light industry achieving its planned share of capital investments by relying more on its own resources. Until the price reforms of 1948, prices of most 'A' sector output were set much lower than costs. As a result heavy industrial enterprises could not be expected to make profits from which they could finance their own expansion. The funds for this came from the centre in the form of massive Gosbank credits. Although with this price structure light industry was the more profitable, that only meant that it was, first, more liable to pay out funds in turnover taxes (in 1939 87 per cent of turnover tax revenue had come from the people's commissariats in the light industry sector)[20] and, second, that it was expected to manage better without any subsidies from the central coffers. Paradoxically, setting prices for 'B' sector output that were more related to costs of production worked to the disadvantage of that sector.

It was only in January 1948 that prices for heavy industrial goods were raised to anything like the level of production costs and many of these increases were partially reversed over the next two years.

The shortage of capital funds in the light industry sector can thus be attributed in part to the indirect effects of the leadership's policies on administrative structure and pricing. Those effects were, however, *directly* the result of various Ministries and other middle-level organisations diverting funds to suit themselves rather than in order to implement the Politburo's sectoral policy.

When light industrial ministries did expend funds on capital construction they did not always make the best use of them. The statistics on investment cited above refer to funds expended, not to actual new capacity brought into use. In 1946 in particular the Soviet construction sector was not working very efficiently; projects were not being completed on time and estimates of costs of construction (where they even existed!) were often proving very optimistic. Construction trusts were accepting little responsibility for the results of their efforts and were seriously deficient in organising their own labour and material supplies.[21]

There were construction problems in heavy industry in 1946, even in such vital spheres as iron and steel production and coal mining.[22] The problems of light industry, however, continued throughout the plan period. Little more than one-third of the production capacity of pure light industry had been rebuilt as late as 1949, in spite of the expansion of investments in this field in 1946 and 1947.[23] The head of Gosplan's capital construction department, Korobov, attributed this partly to the Ministry of Light Industry spreading its scarce construction materials and labour over too many projects. The Ministry had begun to build no less than forty-four new shoe factories over the previous three or more years. Yet by January 1947 none had been completed although a quarter of the allocated funds had been used up.[24] The head of a shoe factory in Kalinin wrote to *Izvestiya* in 1948 that the Ministry had failed to send him any *plans* for the inputs that might have enabled him to complete the construction work required at his plant, let alone the inputs themselves![25]

The 1947 plan for installation of new capacity at seven large textile plants was underfulfilled by more than 60 per cent. However, this was due not just to poor utilisation of materials and labour, but also to a general shortage of such inputs to light industry.[26] The situation in 1949 does not seem to have markedly improved since the Council of Ministers' December 1946 criticism of poor supplies of labour, fuel and power, and new materials to both construction projects and plants in production in the light industrial sector.[27] Even where new machinery was provided or old machines had been renovated in the textile industry they often remained idle for lack of skilled workers and raw materials.[28]

The utilisation of capital expenditures by the pure light industrial sector in particular was far from satisfactory. To what extent was this due to discrimination against this sector by other ministries and organisations who were supposed to supply machinery, materials and labour to it?

The Labour Market

Labour had not unnaturally been diverted away from light industry during the war. The size of the industrial labour force as a whole fell by 18 per cent over the period 1940–5.[29] The number of manual workers employed by the people's commissariats of the Light and Textile Industries fell by 54.9 and 5.1 per cent respectively; the number employed by those of Heavy Machine-Building and the Machine Tool Industry fell by only 18.4 and 5 per cent respectively.[30] There had been no significant improvement by 1947 when light industry still had less than half of its pre-war complement of workers, in spite of the provisions of the December 1946 decree.[31] This had demanded the recruitment of 300,000 new workers for the pure light and textiles industries in 1947 and the training of 305,000 for these sectors by the Ministry of Labour Reserves (MTR) over the next four years. The Minister of the Foodstuffs Industry, V. P. Zotov, saw the cadres problem as the 'dominating' one for his branch in 1947.[32] The Finance Ministry was equally concerned with the cadres shortage in the textile industry.[33] In contrast *Pravda* was claiming as early as February 1946 that the largest sector of heavy industry, machine-building and metal-working, had a sufficiently large and skilled labour force.[34]

The distribution of the manual industrial labour force (that is, excluding employees) in 1950 as compared to 1940 is shown in Table 6.7. That labour force was 23 per cent larger in 1950 than in 1940; yet the light and textile sector's manual workforce had still not quite regained its pre-war dimensions. There had been *some* redistribution of labour in favour of the 'B' sector since 1945, but clearly not enough to enable the planned output

TABLE 6.7. Distribution of manual workforce by industrial sectors 1940–50

	1940		*1950*		*1950*
	No. of workers (1000s)	*% of USSR total*	*No. of workers (1000s)*	*% of USSR total*	*at 1940 = 100*
GROUP 'A'	6476	65.0	8794	71.9	136
Coal	436	4.4	733	6.0	168
Electric power	108	1.1	131	1.1	121
Ferrous metals	405	4.1	605	4.9	149
Chemicals	297	3.0	332	2.7	112
Machine-building and Metal-working	2575	25.8	3332	27.3	129
GROUP 'B'	3495	35.1	3432	28.1	98
'Pure' light industry	2334	23.4	2164	17.7	93
Textiles	1161	11.6	1268	10.4	109
ALL INDUSTRY	9971	100	12226	100	123

Source: See Note 35 on p. 165.

increase of 17 per cent per annum to be achieved without significant improvements in the quality of labour or in the quantity and quality of other inputs. The proportion of the manual labour force working in the pure light and textile industries was still lower in 1950 than it had been in 1940. The poor fulfilment of output plans in this sector must have been partly due to this labour shortage.

To the extent that the labour market was effectively controlled from Moscow, that shortage must be attributed to the work of the Ministry of Labour Reserves. This ministry did do *some* training and recruiting for the 'B' sector ministries. Of the 4.5 million workers to be trained by the MTR over the Five-Year Plan period, 250,000 (or 5.6 per cent) were to go to the textile industry; but 28.9 per cent were to go to the fuel and metals industries alone rather than to light industry.[36] Indeed, from July 1947, most of the major sectors of heavy industry were supposed to recruit nearly all of their labour by means of general agreements with the MTR. This decree was specifically designed to improve recruitment to heavy industry and did not apply to the 'B' sector.[37]

The light industry ministries were left to recruit much of their own labour. In practice, however, the poor work of the MTR also left the heavy industrial ministries to do a large part of their own recruiting, as was officially noted in 1951.[38] The shortage of labour in the 'B' sector was thus not so much due to discrimination against it by the MTR, inspired from above or otherwise. It probably owed far more to the greater incentives heavy industrial enterprises could offer potential recruits in what was in effect a relatively free labour market.

In terms of wages, the Five-Year Plan had already called for higher levels for workers in 'decisive' sectors of heavy industry (coal, metals and oil), although there is no clear evidence of how far this aim was realised.[39] As for bonuses, the 'Director's Fund' was reintroduced in December 1946. From 2 to 5 per cent of planned profits and 30 to 45 per cent of over-plan profits went into this fund in leading branches of heavy industry. The comparable percentages for light industry were only 1 and 15.[40] These funds were partly used to pay bonuses and improve workers' living conditions, but probably had only a marginal impact (many firms, after all, made no profits!). The key fact was, however, that centrally determined wage and bonus levels were often ignored in the drive to fulfil the quantitative plan. Directors of heavy industry plants generally had more resources to attract labour than did their light industry rivals.

For example, light industry certainly found it difficult to provide adequate living conditions for potential recruits. A deputy secretary of Moscow gorkom attacked the local enterprises of the Ministry of the Textile Industry for insufficient attention to the housing problem in 1946.[41] Such complaints were also common in such branches of heavy industry as metals, oil and machine-construction.[42] There was, however, an especially severe problem of high labour turnover in light industry at least in 1946.[43]

Generally it was the factories of those ministries that were best supplied with the funds and the construction materials and the organisations to pay high wages and build new housing that could attract the greater number of workers. They tended not to lie in the light industrial sector. The activities of central ministries did little to help light industry meet its need for large numbers of skilled workers.

Material Supplies and Construction Work

The problems resulting from shortage of labour in sector 'B' were not compensated for by increases in labour productivity. As Table 6.8 shows, output per worker in pure light industry rose by only 22 per cent over the Five-Year Plan period, well below the average figure for all industries and well below the planned average increase of 36 per cent. The major heavy

TABLE 6.8. Increases in labour productivity 1945–50 by industrial sectors

	Increase in gross output per worker 1945–50 (per cent)
All industry	37
Coal	36
Oil	50
Chemicals	76
Machine-building	69
Cement	22
'Pure' light industry	22

Source: See Note 44 on p. 165.

industrial sectors, in contrast, achieved increases of near to or well above the average.

These variations in improvements in labour productivity were only partly due to variations in the quality of the labour force. They were mainly due to variations in supplies of machinery, equipment, raw materials and power with which labour had to work, and probably to variations in the efficiency with which labour was organised and industries administered.

The supplies of machinery to light industry were by no means satisfactory. The plan had decreed that the output of textile machinery in 1950 should be no less than four times its 1940 level. Comparable targets for cutting and pressing machinery and equipment for producing electric power were both set at 'only' 2.5 times 1940 levels. In practice, however, the targets for deliveries of machines to the textile sector were substantially *underfulfilled*, whilst the output target for the whole machine-building sector was *overfulfilled* (by 17 per cent). For example, of the 830,000 machine-driven spinning spindles and 16,879 looms that were to be established in seven large textile enterprises in 1949 only 285,000 and 5567 (respectively) were actually installed.[45] The 1950 output target for weaving machinery was on 35 per cent fulfilled (see Table 6.1). The output of machinery used by many branches of heavy industry (such as metals and electric power) exceeded the plan's targets in 1950. Even the output of agricultural machinery (long the poor relation of the machine-building sector) was above its planned level in 1950. There were shortfalls in some branches of the machine-building industry related to transport and heavy industry. For example, the output of hydro-turbines in 1950 was only one-third of the planned level and that of most railway locomotives and rolling stock only 35 to 45 per cent. The production of metal-cutting and forging tools and of equipment for the oil industry was also lagging behind demand in 1950, although improvements in many of these sectors were noted in 1949–50.[46]

The main criticisms of the Ministry of Machine-Building and Instrument Manufacture (MMP) in 1946, were, however, for its poor performance in supplying the textile industry, not heavy industry.[47] A Council of Ministers decree of September 1946 attacked this ministry for not producing many of the new machines that were required for the reconstruction of the textile industry. The ministry was called on to pay more attention to the needs of light industry.[48]

The machine-building ministries could discriminate between industrial branches in the fulfilment of delivery targets and still fulfil their overall targets (which were based on the value of the machinery produced rather than simply on quantity). The Ministry of Heavy Machine-Building legally controlled the distribution of more than 30 per cent of its output in the form of unfunded production.[49] It seems clear that the MMP had discriminated against the textile sector in 1946 and that other machine-building ministries were little better in this respect.

The productivity of textile machinery in 1950 was in most cases below or very little above its 1940 levels.[50] This must have been mainly due to shortages of labour and materials but also owed something to the poor quality of much of the machinery supplied and the lack of spare parts for it.[51] Machine-building ministries could discriminate against particular sectors in this respect as well as in the simple quantity of machinery delivered. For example, a secretary of the Crimean obkom complained of poor work in the field of repairing and constructing new vessels for the local fishing fleet in 1947.[52] This was (presumably) the responsibility of the recently established Ministry of Transport Machine Building. The same party official pointed to poor supplies of construction materials and construction equipment as a major difficulty for the fishing industry. This he blamed on the lack of specialist (*podryadnye*) construction trusts under the Ministry of the Fishing Industry of the Western Regions. It could also be explained as part of a more general discrimination in the supply of material inputs against light and food industry enterprises by other ministries. Why did this party official attack the fisheries ministries for not establishing specialist construction organisations and not ensuring adequate supplies of construction materials when there were specialist ministries for construction and for the production of construction materials? The answer might well be that little help could be expected in this direction from those specialist Ministries.

Light industry was adequately supplied only with those inputs that were peculiar to that sector and could not be diverted elsewhere. The raw materials for light industry came principally from the agricultural sector, whilst those for heavy industry came mainly from extractive industries (coal, oil, iron ore and the like). The extractive branches mostly overfulfilled their output plans in 1950, the only exception being iron ore output, which was only 1 per cent short of its target. However, the agricultural sector seems also to have fulfilled most of its plans. Those plans were set very low in the first instance and, as is well known, Soviet agriculture had severe problems in this period.[53] Even so, based on the 'biological yield' figures used in the plan, the target for grain output in 1950 was 98 per cent and that for raw cotton 121 per cent fulfilled. Raw material supplies to bakeries and cotton textile plants at least cannot be blamed for the poor performance of the light industry sector. There were, however, problems with the quality of cotton and leather supplied to the clothing and footwear sectors and a shortage of some less widely used raw materials (such as flax).[54]

The main problem for 'B' sector industries lay in poor supplies of those inputs that could be directed towards other sectors. We have already noted the shortfalls in supplies of construction work, machinery and labour to light industry. There is also evidence that fuel and power, transport services and other materials vital to the production process were in especially short supply in the 'B' sector. As a Soviet economic historian of the Khrushchev era noted of the early post-war years:

> local government organs, planning organs and many managers of other branches of industry incorrectly regarded the light and textile industries as of secondary importance and paid little attention to ensuring uninterrupted supplies of fuel, electric power, various auxiliary materials and spare parts to them.[55]

Perhaps the best examples of this phenomenon are the inputs to the textile industry from the Ministry of the Chemical Industry and the supplies of fuel and power to various sectors.

The chemicals ministry and its enterprises were roundly criticised in 1947 for their failure to supply sufficient dyes and new synthetic materials to the textile industry.[56] The obkom secretary of Ivanovo (a major centre of the Soviet textile industry) criticised the same ministry in 1949 for failing to meet the demands of textile plants for dyes, in terms of quality and assortment as well as quantity.[57] Yet the chemical industry as a whole overfulfilled its production target for 1950.

The same secondary priority was accorded to light industry in the supply of electric power. The decree of December 1946 had noted a general shortage of fuel and power in pure light industry.[58] The finance ministry suggested that fuel supplies to the 'B' sector were in need of improvement after the war.[59] Table 6.9 shows that the proportion of electric power used in the production process in this sector declined from 6.3 per cent to 4.8 per cent in 1950. Again the lack of priority accorded to light industry in the base year of 1940 strengthens the argument that electricity supplies to light industrial enterprises did not improve much during the Fourth Five-Year Plan period. Of the 'A' sector industries only iron and steel's share of electric power supplies fell, and this sector's more important source of power was, of course, coal rather than electricity (it was to consume a quarter of Soviet coal output under the Five-Year Plan).

In the provision of material inputs for which it had to compete with heavy industry, therefore, light industry seems generally to have come off second best, as it also did in the provision of capital, labour and machinery.

TABLE 6.9. Proportion of electric power capacity utilised in the industrial production process by sectors 1940–50

	per cent of USSR electric power capacity used in:	
	1940	*1950*
Ferrous metals	18.8	15.7
Coal	9.7	10.7
Oil processing	0.9	0.9
Oil extraction	3.0	3.8
Chemicals	9.3	7.7
Machine-building and metal working	36.1	39.7
Forestry	0.3	0.9
Construction materials	4.8	5.0
Paper and wood processing	5.4	5.2
'Pure' light industry	6.3	4.8
Foodstuffs	5.4	5.5
ALL INDUSTRY	100	100

Source: See Note 60 on p. 166.

Transport

As was previously noted, transport was a severe bottleneck in the Soviet economy in the post-war years. Complaints abounded that light industrial plants in particular were not being properly serviced by the organs of the Ministry of Communications. For example, the director of a salt-producing plant complained to *Izvestiya* in September 1947 that serious underfulfilment of transportation plans was preventing the plant from getting its output to its consumers. The Ministries of Communications and of the Food Industry in Moscow had issued orders calling for improvements in this area but little had been done. 'Why', asked the director, '. . . do they allow their subordinates to ignore government directives and ministerial orders in the Ministry of Communications?'[61] Some five years later, at the 19th Congress, an obkom secretary of the textile-producing Ivanovo region opined that 'There would be considerably fewer shortcomings in the work of Ivanovo enterprises if certain ministries would manage their enterprises more efficiently', and attacked Beshchev, the Minister of Railways, by name.[62] The proportion of railway freight accounted for by the output of major heavy industrial sectors (coal and coke, oil, ferrous metals, timber, ores, firewood and mineral construction materials) and grain rose significantly over the period 1945–50, having fallen slightly during the war period. The share of 'others sectors' (presumably mostly light industry) in consumption of railway transport services fell from 21.8 per cent over the period 1938–40 to 17.2 per cent in 1949–50 (by tons originated), and from 26.7 to 23.4 per cent (by ton-kilometres); the 1945 figures were 21.9 and 31.5 per cent respectively.[63]

In the provision of transport services, as of other inputs, light industry was probably better supplied in *absolute* terms in 1950 than in 1940 or in 1945, but in proportion to heavy industry and to the extra requirements laid on it by the Five-Year Plan, it was clearly lagging. The complaints cited in this section place a large part of the responsibility for this situation at the door of the supplying ministries and their local organs.

Administration

The relative shortage of investment, machinery, labour, material-technical supplies and transport services in the light industry sector was to an important extent due to the practice of those ministries and other supplying organisations of treating light industry as of secondary importance to the leading branches of heavy industry. A recent Soviet student of the period has described in light industry 'a shortage of financial, labour and material-technical resources which were directed in the first instance to heavy industrial concerns'.[64] The complaints cited above place the responsibility for this situation mainly on the shoulders of the various ministries of construction, machine-building, labour reserves and transport and other supplying ministries (such as electric power and chemicals).

Why should these ministries accord first priority to heavy rather than light industry? One answer to this question is that many of those ministries were established primarily to service the reconstruction of heavy industry. For example, in the sphere of construction, the former Ministry of Construction was divided into three separate ministries in January 1946. These specialised in the construction of enterprises in heavy industry, the fuel industry and the military sector. No ministry for the construction of light industrial enterprises was established. This task was the combined responsibility of the heavy industry construction ministry and the light industry ministries themselves. The former (as well as the Light and Textile Ministries) was obliged by the December 1946 decree of the Council of Ministers to concentrate 'the necessary quantity of workers and material-technical resources on construction sites in the light and textile industries'.[65] It was also still expected to provide all the necessary men and resources for heavy industry construction sites. Given the shortages of men and materials at the time the ministry could not fulfil both tasks. It was thus put in the position of *choosing* to which task to accord first priority. The choice of heavy industry reflected both the main work of the ministry and the fact that construction projects in heavy industry were significantly larger than those in light industry.[66] It was easier for the ministry and its construction trusts to fulfil their overall quantitative plans by concentrating on larger (and longer-lasting) heavy industrial projects. That they did so is well illustrated by the official criticisms of 'gigantomania' and attacks on 'serious shortcomings in the management of the construction ministries' in 1950.[67] From the point of view of such ministries faced with central demands to meet *all* priorities, they chose to fulfil those that looked best in terms of the quantity of work they had done. (That quantity was measured in roubles expended on construction rather than the amount of new productive capacity established; the Five-Year Plan was seriously underfulfilled in all sectors in this latter respect.) To that extent the Soviet administrative system of 1946–50 was ill-adapted to the task of expanding construction in light industry.

The arguments concerning construction can equally be applied to the machine-building sector. For most of the Fourth Five-Year Plan period there existed six specialist machine-building ministries, but none were specifically designed to meet light industry's needs. The transport, construction and agricultural sectors all had specialist machine-building ministries servicing them. The other relevant ministries specialised in heavy machine-building, machine tool production and the manufacture of motors and instruments. Like most of the other ministries in charge of material-technical supply, these three supplied both light and heavy industry. Partly as a result of this light industry received only meagre supplies of new machinery and the 'basic' reconstruction of light industry had to wait until the Fifth Five-Year Plan period;[68] the 'basic' reconstruction of heavy industry in contrast was completed at the latest estimate by 1949.

Sectoral policy outcomes were only indirectly the result of leadership decisions on the ministerial structure. A number of these specialist servicing ministries were established in 1945–6, but there is no evidence that the leadership realised the impact their creation might have on light industry. That impact was *directly* the result of decisions taken within ministries and not by the leadership.

Policy outcomes might therefore be explained by the fact that these ministries and their enterprises could better fulfil their overall quantitative plan targets by constructing and producing machinery and supplies for heavy rather than light industry. The ministries' behaviour might also be explained in terms of the habits of their officials. Perhaps the pre-war and wartime habits of giving some parts of heavy industry sectors absolute priority over all other sectors could not simply be altered by decree from above after the conflict. Hardt and Frankel argue that the 1930s Stalinist system of mobilisation of resources to meet primary goals (the expansion of basic branches of heavy industry) determined the character of the 'groups' of industrial (enterprise) managers for the whole Stalin era.[69] In terms of social background, experience and the like, these managers formed an increasingly homogenous group from the late 1930s and their tenure increased dramatically from then on (from an average of three years in the 1930s to one of ten years or more in 1953). These authors go on to detail the means by which these enterprise directors gained cohesion as a group and influenced their bureaucratic superiors in the ministries in directions that did not fit in with the latter's own interests (such as enterprise autonomy). Such influence within a ministry was only to be expected in the matter of according priority to certain sectors of heavy industry. In this case the interests of both ministry and enterprise officials (in quantitative plan fulfilment) were broadly similar; and many of the heads of the chief administrations (glavki) of the ministries had been recruited from amongst the ranks of the enterprise directors. By the late 1930s some two-thirds of glavk heads had previously been employed in administrative positions (*sluzhashchie*).

Whether the officials' behaviour was more the product of their previous career background or of the interests derived from their current occupation must remain an open question. In many cases these two factors were mutually reinforcing. What the evidence presented here shows is that officials did often discriminate against light industry *in practice*.

The 'B' sector ministries themselves naturally had an interest in enforcing the priorities of the Five-Year Plan and the December 1946 decree. Not only did they lack the resources to do so; they did not always make the best use of the resources they did get. Criticisms of the poor work of the textile ministry were one of the main reasons for its reincorporation into the pure light industry ministry in December 1948. For example, the ministry was criticised at a joint plenum of the Moscow *gorkom* and *obkom* (city and province party committees) for meddling in the affairs of enterprises without having any clear idea of what it was doing.[70] Both the textile and light industry ministries were slated for their poor utilisation of machinery in 1946.[71] The official criticisms of them after their amalgamation were that they had been too concerned with quantitative plan fulfilment rather than with producing saleable goods, and that, in the case of the textile sector, plants tended to be too specialised (and thus, presumably, too reliant on outside suppliers!).[72] Similar criticisms were, however, also made of some heavy industry ministries.[73]

Indirect Controls over Sectoral Policy: The Checking Organs

Any enforcement of the 1946 decisions on sectoral policy thus depended on the power of the checking organs to seek out and redress the biases of the ministerial framework that formed the heart of the command economy.

The Party

In theory the most powerful arm of the Soviet Politburo, the party in practice often proved all too ineffective as an enforcer of economic policy. Party organs were often criticised for their lack of control over the economy in the years after the war. For example, in March 1946, B. N. Chernousov, a secretary of the Moscow obkom, attacked local party organs for poor supervision (*kontrol*) over production matters in the textile industry.[74] Later in the same year a *Pravda* editorial called for improvements in cadres work within central ministries and economic agencies.[75] This criticism was directed primarily against the Cadres Directorate of the Central Committee Secretariat. In 1946 some elements within the Party were concerned that the apparat was failing in its task of checking on the implementation of economic policy both at the level of local party organs' involvement in day-to-day production matters and at that of personnel selection in the state organisations directing the economy.

The reform of the Central Committee Secretariat along production branch lines in 1948 was designed to strengthen central party supervision over economic affairs. No doubt the reforms also owed something to the death of the ideologically minded Zhdanov and the rise of the more production-oriented Malenkov within the Secretariat.[76] Their significance for the present argument, however, lies in the fact that the functions of the old Cadres Directorate were distributed amongst departments specialising in particular industrial branches, amongst them being separate departments for heavy and light industry. Such specialised agencies would, in theory at least, be more able to relate cadres policy to the particular problems of each industrial branch. Where leadership policy was not being fully implemented these departments were to seek out the reasons and take appropriate measures. They could therefore have helped to implement sectoral policy by means of purges of ministerial officials. In practice, however, there is no evidence of any purges at the time associated with sectoral policy. This is not very surprising when one considers that light and heavy industry were supervised by separate departments. Only the Planning, Trade and Finance Department of the Central Committee Secretariat had any overall responsibility for sectoral policy, but it lacked the production specialism and links with branch ministries that the other two departments had. The weakness and parochialism of many party officials was summed up by Malenkov's 1952 complaint that many party organs were

> incapable of meeting all sorts of . . . narrowly departmental . . . pulls and pressures, and are overlooking outright distortions of the Party's policy in economic fields and violations of the interests of the state.[77]

This was due partly to the fact that the Central Committee Secretariat was itself organised along departmental lines. It was also due to the general weakness of local party organs at this time. Party *kontrol* over industry was exercised through local organs' relations with enterprises and ministries as well as through Secretariat pressure on economic agencies in Moscow. The weakness of these local organs has been noted in Chapter 2. Their lack of expert knowledge was not helped by the abolition of the post of deputy secretary for particular industrial branches at obkom, gorkom and *raikom* (district party committee) level in 1948. At these levels a united industry department was to be established. In many cases these departments of local secretariats acted as allies of local government agencies rather than as the eyes of the Central Committee against them. For example, A. P. Yefimov, secretary of Khabarovsk Krai, in accusing the local heavy industry department of being little more than a 'despatcher's office' helping state enterprises find scarce supplies, called on them to pay more attention to 'party political work', in other words to cadres, checking and political education work.[78] At both local and All-Union levels the party apparatus in the period 1945–53 was ill-equipped to ensure the fulfilment of the leadership's sectoral policy against the wishes of the heavy industrial and supplying ministries. It lacked the authority and often the will to do so.

Checking Agencies within the Council of Ministers

The main agency for co-ordinating the work of the various economic ministries (apart from the Council of Ministers itself) was Gosplan. Its co-ordinating role became ever more important after the war as the number of separate specialised ministries expanded. The responsibility for securing the planned proportions in the Soviet economy was thus greater than ever after 1945.[79]

Within Gosplan summary plans for transactions between branches of industry (and thus between ministries) were from 1946 the responsibility of the Production Department. This department, however, had separate sectors for heavy and light industry. Gosplan was as liable as the ministerial parts of the Council of Ministers to the evils of departmentalism (*vedomstvennost*) in its implementation of sectoral policy. To overcome such problems Gosplan was to make particular use of the system of 'planning by balances'.[80] Gosplan and the newly established Ministries of Food Reserves and of Material Reserves were to ensure that all branches were well supplied with the inputs they required, whether from other ministries or from the state reserves that were set up.

Gosplan's departmentalist tendencies have already been noted, however, and in December 1947 many of its functions in this field were transferred to Gossnab. In July 1948 the Ministries of Food and Material Reserves were merged and in 1951 converted into an agency of the Council of Ministers (*Gosprodsnab*). The impact of these changes on sectoral proportions and co-ordination between ministries seems to have been minimal. Such frequent administrative reorganisations were a sign that the organisation concerned was simply not carrying out its tasks effectively. In so far as these supply-co-ordinating agencies were organised along the same departmental lines as the ministries themselves and lacked the expertise and number of staff of the ministries, they could not provide a more effective alternative channel for the implementation of sectoral policy.

Of the other checking organs within the Council of Ministers, the secret police had little noticeable impact. There is no evidence of mass purges of ministerial officials in any sector of Soviet industry after the war.

The extent to which financial controls could be used to enforce sectoral policy was severely limited by the price structure for industrial goods. If an enterprise's and ministry's future investments and supplies were to depend on its making profits and fulfilling its obligations to other organisations, then it had to be given some prospect of making profits and some punishment for not fulfilling its obligations. Yet prices for most heavy industrial goods were set so low as to preclude profit-making and thus much financial accountability. The state subsidy to such enterprises rose from 13.9 billion roubles in 1945 to 35.3 billion roubles in 1948 (about one-fifth of GNP at that time).[81] It was only after the price increases of 1949 that the subsidy fell to 2.9 billion roubles in that year. Only then did financial accountability and the use of financial levers over ministries become possible. Even then, the old practices of setting prices for new lines of output at levels reflecting narrow departmental interests and even ignoring the centrally fixed prices for output altogether continued.[82]

In theory firms were liable to heavy fines if they failed to fulfil their contractual obligations to other organisations consuming their output. In conditions of general scarcity, however, no firm could hope to meet all its contracts on time. Most therefore continued to put enterprises within their own ministries first,[83] and often simply left any fines that accrued to them unpaid! In fact the damaged enterprise would often not try to collect the fine at all for fear of offending the supplying ministry.[84] In practice all the financial and other controlling organs could do was to report errant enterprises and ministries to higher authority. Even within the Council of Ministers such reporting seems to have had relatively little impact on the actual work of ministries.

Other Checking Agencies

The executive competence of local Soviets in economic matters did not extend beyond small-scale local and co-operative industry, social services and housing. Even in the sphere of local and co-operative industry, however, Moscow took a keen interest in the early post-war years. The Council of Ministers decree of 9 November 1946 clearly indicated that the leadership expected a significant part of 'B' sector growth to emanate from locally controlled enterprises, and a special glavk under the All-Union Council of Ministers was established to assist in this task. Union republic ministries for local and co-operative industry were also called upon to give all assistance to local enterprises to produce more consumer goods.

However, local industry was not immune to the heavy industry bias that permeated the central industrial administration. A *Pravda* editorial commentary on the November 1946 decree accused local Soviet and Party organs of not paying enough attention to the problems of local consumer goods industries, the Chuvash and Penza regions being singled out for criticism.[85] A letter published by *Pravda* in October 1949 accused the machine-building glavk of the RSFSR Ministry for Local Industry of forcing two factories producing knives and razor blades to branch out into the production of 'A' sector goods, thereby preventing them from fulfilling their plans for output of mass consumption goods.[86] Local administrations, like their All-Union counterparts, were usually organised along departmental lines and so were just as prone to vedomstvennost and as subject to the power of the glavk. They could be expected to play only a very limited role in implementing the leadership's sectoral policy.

In summary, most of the non-ministerial arms of the political leadership did not possess the power to enforce sectoral policy by bringing supplying agencies to heel. Further, many were themselves organised along departmental lines and so lacked officials who had much interest in reversing the heavy industry bias exhibited by many glavki. Supply glavki became such a law unto themselves that there were reported cases of deliveries that were not requested and not even wanted, as well as the more normal shortages.[87]

Conclusion: Responsibility for Sectoral Policy Outcomes

We have argued that the sectoral policy actually carried out over the period 1946–50 was far more favourable to heavy industry than had been laid down by leadership decrees in 1946. The divergence between policy decision and fulfilment was mainly due to the distortions in the supply of financial and material resources by ministries that other organisations were powerless to correct (and often unwilling to correct).

The priority sector model might explain the divergence between plan and fulfilment as due to the leadership's placing *overriding* emphasis on heavy industry. According to this model, the Politburo might have set its targets for the 'B' sector in good faith, but, if it found that targets for both sectors could not be fulfilled, then light industry would be accorded very little priority. There is, however, no evidence of such unity of purpose amongst the leadership on sectoral policy at this time. In any case the fact that the plan for industrial output as a whole was overfulfilled by 17 per cent suggests that targets for *both* major sectors could have been fulfilled. Indeed that for the 'A' sector was overfulfilled by 28 per cent and that for the 'B' sector underfulfilled by 5 per cent. The reason for light industry's shortage of inputs was not that these were sorely needed to enable heavy industry to meet its target; that it did with ease. An analysis of sectoral policy-making in terms of the priority sector model must therefore fall back on the argument that the leadership's real aim was to *overfulfil* the target for heavy industry in 1950. The proposition that the Five-Year Plan and subsequent decrees were mere window-dressing has been dealt with elsewhere in this book, as has the argument that the leadership's sectoral priorities were substantially altered during the Fourth Five-Year Plan period. There is no evidence to show that leadership priorities were in practice so different from those publicly announced in April 1946 or that they were ever substantially altered in favour of heavy industry.

Above all, most of the detailed evidence presented in this chapter suggests that the major limitation on the growth of a sector was the supply of inputs to it, and the fact that control over those inputs was in the hands of ministerial officials rather than their nominal masters. As in the case of regional policy it seems that decisions from above were not automatically carried out by the apparatus of the 'command' economy. Even in the Stalin era these officials had to be coaxed into obedience. The only way of 'coaxing' them into paying more attention to the need to develop light industry would presumably have been a combination of re-education of officials at all levels and a fundamental restructuring of the ministerial framework. Only under these conditions would officials of supplying ministries be inclined by their past experience and the interests of their current occupations to carry out the regime's priorities.

7 Conclusion

The general arguments of the earlier chapters and the detailed case studies of the latter half of this book all point to the same conclusion: that the basis of the command economy was *not* an automatic or an unquestioning obedience to commands. According to the simple model of the command system presented in Chapters 1 and 2, Stalin gave orders to his Politburo colleagues who in turn transformed them into more detailed instructions for ministries and other subordinates. In turn the ministries used these instructions as the basis for their own directives and orders which they then gave to factories under their control. The possibility of any distortion of directives in any link in this Stalin-Politburo-ministry-factory chain was obviated by the activities of the checking agencies within and outside the Council of Ministers framework.

The primary concern in this book has not been with the Stalin-Politburo, or the ministry-factory linkages, but with that link between the Politburo and the ministries. Before discussing the latter linkage, however, it should be noted that in both this study and in others[1] it is indicated that the other two linkages were not simply commands. In the cases of these two linkages the direction was upward as well as downward and the links themselves were often far weaker than is suggested by the simple model of the command economy. Stalin's relations with his Politburo in the post-war years were not those of a king and his courtiers. As we have seen in Chapters 1, 3 and 5, Politburo members such as Molotov, Malenkov, Zhdanov and Khrushchev argued with one another over vital policy matters, over regional and sectoral policy, as well as foreign affairs and agriculture. This is not the place in which to argue Stalin's role in all these disputes. Suffice it to say that Stalin, like most dictators, had to pay some heed to his colleagues' wishes and that those colleagues had their own power bases in the bureaucracy. The General Secretary himself could not always manipulate these bases. For example, Malenkov's support amongst the senior officials of the industrial ministries put him in a very strong position. Some senior Politburo members did influence Stalin's decisions, and, in some spheres, especially those involving domestic policy, they were allowed to make their own decisions uninfluenced by Stalin. Roy Medvedev argues that in the 1940s Stalin began to exhibit a general neglect of affairs of state and that after the war he 'no longer went into the "petty details" of the economy'.[2] Alexander Werth talks of Stalin governing mostly by proxy for the last five years or so of his life. This was partly due to the General Secretary's deteriorating state of health. He was seventy in 1949 and had made no public speeches between 1946 and 1952. His ten-minute speech to the 19th Congress in the latter year was clearly a great effort for him. Stalin therefore allowed his colleagues some freedom of action and in that sense the linkage between the two was often weak. But the linkage was sometimes in an upward direction, with factions within the Politburo seeking to influence Stalin. At the level of ministry-enterprise relations, factory managers were able to exert some influence over their ministerial superiors. Factory directors could influence the command which ministries gave to them: both in the formal process of plan formation and by the informal practices of hoarding, falsification and so on. This linkage was again a two-way process and in its downward guise (ministerial orders to plants) it was often feeble. Many managers seemed to spend more time hoodwinking their superiors and engaging in semi-illegal behaviour than in actually running their factories according to the plan. Yet in a very real sense the factory manager did need the ministry's goodwill, for it was this body that had operational control over many of the scarce supplies which plants needed in order to survive.

The focus of this book has been on the Politburo-ministry linkage. Chapters 3 and 5 have shown in detail how officials of the command economy outside the Politburo sought to influence the leadership's decisions and plans. Ministers, planners and their subordinates joined in public debate with academic advisers in specialist journals and at various conferences and meetings on both the sectoral and regional issues. They were also able to exert more private influence over the process of plan formation as they provided the information on which final decisions were to be based. Of course the leadership might have decided that a degree of discussion beforehand would enable it to produce better plans. Yet the discussions continued after the approval of the Five-Year Plans in 1946 and 1952. The fact that the Politburo had reached a decision did not stop officials and academics continuing to try to change or reinforce that decision.

They continued to discuss issues of sectoral and regional policy because such debate clearly *did* influence decisions in these spheres. Analysis of the decisions reached, especially in Five-Year Plans, shows that they were indeed compromises between differing points of view that had been expressed both within and outside the ranks of the Politburo. The Old West and the 'B' sector were, according to the Fourth Five-Year Plan, to regain some, but not all, of the ground they had lost to the Near East and to heavy industry respectively during the war.

Chapters 4 and 6 show that in the course of their implementation these decisions were distorted in favour of the industrial areas of the Old West and the 'A' sector. In the case of regional policy, production-branch ministries had little interest in ensuring that regional proportions were in accordance with the plan. Their major concern, as was argued in Chapter 2, was the expansion of their 'empires' and the survival of the ministry. To achieve these aims they had to direct resources to those regions where the yields on investment were greater in the short run. These were the regions of the Old West and the Baltic States, not the eastern regions where so much growth had been planned. To divert resources in this way ministries had to make use of their greatest power, which was their control over physical appropriations and expenditures. In exercising this power they exploited both their legal rights and the inability of the checking agencies to keep them to the letter of the law and the letter of the plan. In pursuing the aims of survival and expansion the branch ministries therefore achieved their third objective, that of extending their own autonomy.

In the case of sectoral priorities some ministries had an interest in carrying out the leadership's policy on the expansion of consumer goods. Some clearly did not. In the area of sectoral policy the behaviour of the ministries which supplied both heavy and light industry was crucial. If the ministries of machine-building, construction, transport, and others, had been willing to enforce the sectoral balance announced in 1946, only then could it have been achieved. They proved themselves unwilling by diverting supplies to the heavy industrial ministries. Their motives for doing so were probably, first, sheer force of habit and training; the idea of primacy to heavy industry had been almost an article of faith in the 1930s and officials trained in that era were not about to abandon it merely because the leadership asked them to do so. Their second motive lay in the bargaining power of heavy industrial ministries. The Ministry of Ferrous Metallurgy, for example, was far more able to provide favours (again, mostly scarce supplies) for the Ministry of Transport than was the Ministry of the Textile Industry. The former provided the iron and steel without which railways could not be run; the latter provided only uniforms. Much the same could be argued of the other supplying ministries. They pursued their own development, not that of light industry as prescribed by the leadership.

The complexity, parallelism and sheer size of the command economy structure thus allowed officials within it a significant freedom to make their own decisions. As established, experienced and highly trained men they often utilised their freedom of choice to put their own interests above those of the country as a whole, as defined by the Politburo. Whatever the leadership resolved even under the influence of some of its subordinates, the ministries did what they decided was in their own best interests. This book has shown how the formal structure of the command economy allowed them to do this. It has shown how the actual performance of the economy in both its sectoral and regional aspects reflected branch ministerial interests. Numerous complaints about ministers and their officials acting in this way have also been cited. In judicial terms evidence has been produced of the opportunity and motives to commit such crimes. The 'crime' being not carrying out the leadership's orders, the criminals were the ministries and also other organisations which could not or would not check on the guilty ministries. Evidence of such crimes being committed has been given. Yet there was no punishment. Short of a purge similar in scale to that of 1936–8, Stalin could have done little to restore his control over the machine that he, above all others, had created and set in motion. Perhaps it was just such another purge that he was preparing at the time of his death.[3]

That ministries did not always do as they were told is of little surprise to any student of the command economy. But many specialists in this field cling to the idea that in major priority sectors the leadership could and did get its own way. Yet the command economy after the war did not prove an adequate machine for attaining a 'limited number of well defined objectives'. Sectoral and regional balances were clearly defined in the Fourth Five-Year Plan and yet were not attained. Nor was the extent of the distortion marginal; output target fulfilment varied by 20 or 30 per cent between main sectors and regions; and by much more in individual industrial branches and economic regions. The command economy could not be used for achieving even a few clear objectives.

Of course the basis of the priority sector model is the assumption that all the leadership wanted was to increase the output of a few sectors of heavy industry as fast as possible and that anything else was a purely secondary priority. The analysis presented in this book of the policy decisions announced by them and apparently transmitted in all good faith to the officials of the command economy provides little foundation for this assumption. In any case the policy-formation process tended to produce decisions that did not simply adhere to this one absolute priority. The compromise decisions it produced were not those of simple priority to iron and steel and machine-building, and they were of a sufficient complexity (although basically clear enough) to put them beyond the reach of leadership achievement. The leadership's priorities were only fulfilled if and when they were also the priorities of the relevant parts of the bureaucracy. Heavy industry did grow rapidly after the war. The evidence of this book suggests that it did so *in spite of* the leadership's wishes and not because of them.

Alfred Meyer has drawn an analogy between the Soviet system and a giant western corporation.[4] This ideal of a centralised and disciplined 'USSR Inc.' assumes the simple vision of the command economy outlined in the first section of Chapter 2. A more apt analogy would be between the ministries of the Stalin era and Western businesses. Relations between Soviet ministerial empires correspond far more closely to the bargaining relationships of an imperfectly competitive market than they do to the commands which Meyer thought characterised the internal workings of American corporations. Ministries found it far easier to get the supplies they needed by trading goods and influence with other ministries and enterprises than by making complaints through the formal paths of the Council of Ministers and other checking organisations. Ministries operated in their own interests—to ensure their own survival, expand their own empires and extend their own autonomy,[5] and it seems that there was very little that the political leadership could do. It was an American President, Harry S. Truman, who once said that if General Eisenhower became President in 1952, 'He'll sit here and he'll say, "Do this!" "Do that!" and nothing will happen. Poor Ike—it won't be a bit like the Army. He'll find it very frustrating.'[6] This lesson was one that Stalin's Politburo had not learned and indeed is one that its present-day successor is still grappling with. To ensure the implementation of an economic policy the Politburo had to either make sure that it accorded with the majority of ministerial interests or change those interests by restructuring the ministries. This is what Khrushchev sought to do in 1957 and Kosygin in 1965 and 1973 and in each case ministerial officials have frustrated the reorganisation itself.[7] Perhaps Stalin had no alternative but to accept the power of departmental interests within the command economy.

The conclusions presented here are based on a study of only two aspects of economic policy during only one period of Stalin's rule. How generalisable are they? It seems reasonable to apply the same reasoning to other spheres of economic policy in the same era, but do our conclusions apply to other policy spheres, to other periods or even to other political systems? Studies of foreign and cultural policy-making in the post-war years suggest that the command economy is not the only sphere of activity in which scholars may have overrated the power of Stalin and his Politburo. Could the power of the ministries be explained by circumstances peculiar to the years 1945–53? The 1936–8 purges and the Second

World War certainly weakened the major checking organs, the party and the police, whilst the war itself showed how indispensable the state industrial machine was. By 1945 many officials of the command economy enjoyed a degree of experience and security that they had not seen in the 1930s. Perhaps they knew that the leadership was well aware that any violent attack on their power might be very damaging to the industrial growth rate and ultimately to the survival of Stalin's regime. In addition to this the position of Stalin himself was weaker after the war than before it. This was partly because his health was declining and he spent more and more time on foreign affairs and writing books on economic theory and linguistics than on deciding matters of domestic policy.

Undoubtedly the circumstances of the late 1940s and early 1950s were much more suitable for the exercise of ministerial muscle in the USSR, but one might conjecture that our conclusions could equally apply to the Soviet Union in the 1930s and indeed to other systems with large and complex bureaucracies. In the absence of systematic and large-scale coercion highly trained and experienced government officials cannot be expected to act as little more than the tools of political leaders. As bureaucratic organisations grow in size and complexity, and as the levels of experience and education of bureaucrats rise, the opportunity as well as the desire to impress their own viewpoints on the policy-making process increases. For example, the influence of highly trained and experienced civil servants over inexperienced Cabinet Ministers in Britain has been extensively documented. Richard Crossman's battle with his Permanent Secretary, Dame Evelyn Sharpe, illustrates a similar phenomenon to Zverev's suggestions to the Council of Ministers on the role of consumer goods output in the Fourth Five-Year Plan period.

Whether a highly educated and extensive bureaucratic structure helps or hinders the execution of a political leadership's decisions is a problem faced by many third world nations today. In some African countries government officials seem more concerned with their own financial welfare than with carrying out the centre's instructions. The Nigerian leader General Murtala Mohammed removed, on charges of corruption, most of the provincial governors of the previous regime. The action of the Soviet ministries in the 1940s is not so very different from such corruption. The general proposition from the study of the post-war Stalinist command economy is this: as the degree of specialism and size of the bureaucracy ('technocracy') increases, so the influence of bureaucrats on policy-formation grows; and, at the same time, the level of policy-implementation declines. It can only be verified by many more studies of bureaucratic behaviour in different political and economic systems.

Appendixes

Appendix A: The Reliability of Soviet Economic Statistics for the Period 1945–53

The vast majority of the data used in this book have been drawn from Soviet statistical handbooks and other official sources (mostly decrees and plan documents). The reliability of such data has been questioned by many Western authors (for example, Jasny, Hodgman and Nutter).[1] There are three obvious reasons for doubt in the case of statistics from the Fourth Five-Year Plan period:

1. That the data must have been collected in the situation of general disorganisation and administrative reconstruction that followed the war. Indeed a separate Central Statistical Administration was established only in August 1948.
2. That there is good evidence that some Soviet data from the time were falsified. The use of 'biological' yields to inflate agricultural output figures has since been officially admitted. There may have also been good reasons for concealing the real extent of Soviet defence expenditures at this (and other) times.
3. That industrial output was still based on 1926–7 prices, which enabled managers to inflate output figures by overpricing new types of product, and did not reflect comparative production costs in 1945–53.

The arguments for the use of these data are both specific and general. Specifically, all of the data here are consistent over time (unless otherwise noted); in other words the same series and the same figures are presented in successive statistical handbooks. Agricultural data for the late-Stalin era *were* revised in later handbooks. Either there was no need to revise most of the industrial data or the same lies are still being told. There is no evidence or obvious reason for the latter being so.

All the output data used in our tables are at 1926–7 prices so are internally consistent (or are measured in physical units). We have been careful to note the prices in which various figures for capital investment have been presented. These two specific points are part of two general arguments for the reliability of these data. First, the data are used only to compare different areas and sectors *within* the USSR. The same 1926–7 prices applied to all areas and were fixed for sectors. In addition the analysis presented here relies on relative *rates of growth* rather than absolute levels of output or investments. The first general argument, therefore, is that these data are adequate for comparisons across time, sectors and areas within the USSR, although they may not be useful for (for example) static comparisons of living standards between nations. If there are biases in the data there is no obvious reason to suspect that they were any more influential at one time or in one area or sector rather than another.[2]

Our second general argument is that the data here appear to have been what Soviet planners, politicians and administrators at the time used. Of course, they had access to more extensive data but the figures produced by Prikhod'ko, Dokuchaev (and others) are consistent with those in the statistical handbooks. The significance of that is in the fact that both authors quote as their sources archives—those of Gosplan in the first case and those of various republics and regions in the second.

Similarly, although the 1926–7 price sets were rather outdated by 1946, they were the ones that were actually used in compiling plans and assessing their fulfilment during our period. In general, therefore, the data used here are adequate for the purposes to which they are put.

Appendix B: The Compilation and Significance of Incremental Capital-Output Ratios

The Incremental Capital-Output Ratios (ICORs) presented in this book were calculated as follows:

$$\text{ICOR} = \frac{\text{Capital expended over period 1946–50 (inclusive)}}{\text{Output in 1950 } \textit{less} \text{ output in 1945}}$$

Thus the higher the ICOR, the more capital had to be (or was planned to be) expended to yield a given increase in final output. The lower the ICOR the greater the returns (in terms of increase in final output) from a given amount of investment over the five-year period. Thus, in terms of our criteria, the lower the ICOR, the greater the emphasis on maximising short-run (up to five-year) returns.

However, problems of data availability did lead to some compromises having to be made in operationalising this measure. The first of these concerns the periodisation. Why should investment over the period 1946–50 be expected to yield returns over the same period rather than, say, over 1951–5? The answer is simply that if it was not expected to, then longer run criteria were being used. In other words, if we define the short run for our purposes as being up to five years, then this ICOR measures the extent to which short- rather than long-run criteria (expectations) characterised both plan targets and their fulfilment.

Second, in the case of regional ICORs, full sets of industrial investment figures for the period were not available. The data actually used are those for all centralised investment (excluding that by collective farms and individuals). This includes investment in the construction, transport, trade and state agricultural sectors. But it may be more valid to include rather than exclude investment in these sectors, for all give returns in the form of industrial output. Investment in building, transport, trade and even agriculture yield supplies of services and goods essential to the growth of industrial output. In any case the same measure was applied to each region and any large inter-regional biases that might have resulted would have been reflected in the differential supplies of various inputs to different areas and industries analysed in Chapters 4 and 6.

The third problem of measurement concerns prices. Industrial output figures are measured in 1926–7 prices and capital investment in 'comparative' (constant) prices apparently calculated in about 1960. To partly overcome this problem only comparative rather than absolute ICORs have been used. In general, we must re-emphasise the point made in Appendix A that the data are adequate for comparative purposes within the USSR, for comparing different regions, sectors and times and (especially) for comparing plan targets with their fulfilment. Finally, there is some justification in using whatever data are available, provided that their limitations are clearly noted, and especially if they reinforce other (non-statistical) evidence.

Notes

The following abbreviations are used in the notes which follow:

IVOVSS: Istoriya Velikoi Otechestvennoi Voiny Sovetskogo Soyuza
PKh: Planovoe Khozyaistov (journal of Gosplan)
Resheniya: Resheniya Partii i Pravitel'stva po Khozgaistvennym Voprosam
SEOV: Sovetskaya Ekonomika v Period Otechestvennoi Voiny
VE: Voprosy Ekonomiki (journal of the Institute of Economics)
VOSS. Ind: Vosstanovlenie Industrii

Full titles of articles in these and other journals have been omitted to ensure clarity. Thus the reference A. Zelenovskii, PKh, 1946. no. 1 is to an article by Zelenovskii in *Planovoe Khozyaistvo*, no. 1 for 1946.

1 Introduction

1 N. Jasny, *Soviet Industrialisation 1928–52* (Chicago University Press, 1961) p. 235.
2 R. Conquest, *Power and Policy in the USSR* (London: Macmillan, 1961).
3 M. Djilas, *Conversations with Stalin* (London: Hart-Davis, 1962) p. 134.
4 L. Schapiro, *The Communist Party of the Soviet Union* (London: Methuen, 1963) pp. 507–8.
5 W. O. McCagg, Jr, *Stalin Embattled 1943–8* (Detroit: Wayne State University Press, 1978).
6 M. Shulman, *Stalin's Foreign Policy Reappraised* (Cambridge, Massachussetts: Harvard University Press, 1963).
7 See McCagg, pp. 249-54.
8 See E. Frankel, 'Literary Policy under Stalin in Retrospect: A Case Study 1952–3', in J. P. Shapiro and P. T. Potichnyj (eds), *Change and Adaptation in Soviet and East European Politics* (New York: Praeger, 1976); V. S. Dunham, *In Stalin's Time: Middle Class Values in Soviet Fiction* (Cambridge University Press, 1976).
9 McCagg, pp. 304–5.
10 *Khrushchev Remembers*, vol. I (London: Sphere Books, 1971) p. 555.
11 McCagg, p. 307. The Mingrelian Affair was a purge of many Beria supporters in a district of Georgia in 1951–2.
12 Jasny, pp. 242–3.
13 For example, see A. G. Frank, 'The Organisation of Economic Activity in the Soviet Union', *Weltwirtschaftliches Archiv 1957*, pp. 104–8.

2 The Organisation of Control in the Command Economy

1 H. A. Simon, 'Administrative Behaviour', *Encyclopedia of the Social Sciences*, vol. II, pp. 77–9. He argues that the bureaucrat's behaviour is 'bounded' in its rationality by the specialism of his sub-section of the hierarchy (each sub-section having its own 'sub-goals' and its own sources of information which affect its perceptions of reality) and by the need to choose a reasonable alternative rather than seeking out the best possible solution to a problem ('satisficing').
2 This account is based on that given by J. Miller, 'Soviet Planners in 1936–', in J. Degras and A. Nove (eds), *Soviet Planning: Essays in Honour of Naum Jasny* (Oxford: Blackwell, 1964) p. 126.
3 From here onwards Miller's account is modified in accordance with a Council of Ministers decree of 28 August 1946, in *Resheniya Partii i Pravitel'stva po khozyaistvennym voprosam* (hereafter *Resheniya*), tom 3 (Moscow: Politizdat, 1968) pp. 334–5. This decree really altered only the timing of the process. The drawing up of Five-Year Plans followed a broadly similar pattern.
4 A. G. Frank, p. 116.
5 For example see A. Arakelyan, *Industrial Management in the USSR*, translated by E. L. Raymond (Washington D.C.: Public Affairs Press, 1950) p. 140; and *Ekonomicheskaya Istoriya SSSR*, pod. red. I. S. Golubnichogo *et al* (Moscow: Mysl', 1967) p. 440.
6 Arakelyan, p. 140.
7 *Resheniya*, tom 3, pp. 334–5.
8 *Organizatsiya Finantsirovaniya i Kreditovaniya Kapital'nykh Vlozhenii*' pod. red. N. N. Rovinskogo (Moscow: Gosfinizdat, 1951) pp. 50–2.
9 *Resheniya*, tom 3, p. 335.
10 These reforms raised the prices of many heavy industrial products above their production costs. Only after this could heavy industrial enterprises make profits to finance their own investments.
11 *Resheniya*, tom 3, p. 428.
12 G. P. Kosyachenko, *PKh*, 1946, no. 2, p. 144.
13 A. M. Rubin, *Organizatsiya Upravleniya Promyshlennostyu v SSSR (1917/1967 gg.)* (Moscow: Ekonomika, 1969) p. 147.
14 G. P. Kosyachenko, *PKh*, 1946, no. 4, p. 13.
15 Frank, p. 166.
16 Ibid., p. 128.
17 D. Granick, *Management of the Industrial Firm in the USSR*, (Columbia University Press, 1954) p. 75.
18 For example, *Pravda* 16 July 1946, p. 1 (editorial) and 31 July 1946, p. 1 (editorial).
19 V. Grossman, *PKh*, 1946, no. 3, p. 37 and pp. 29–40.
20 *Resheniya*, tom 3, p. 612.
21 Ibid., pp. 674–7.
22 For a summary see B. D. Wolfe, *An Ideology in Power* (London: Allen & Unwin, 1969) part IV, section 1.
23 For example, see Yu. A. Prikhod'ko, *Vosstanovlenie Industrii*, 1942–50 (Moscow: Mysl', 1973) (hereafter *Voss. Ind.*) p. 189; and G. A. Dokuchaev,

Rabochii Klass Sibiri i Dal'nego Vostoka v Poslevoennye Gody, 1946–50 (Novosibirsk: Nauka, 1972) p. 35.

24 See Frank, pp. 137–8.
25 Prikhod'ko, *Voss. Ind.*, p. 214.
26 *Razvitie Sotsialisticheskoi Ekonomiki SSSR v Poslevoennyi Period*, pod. red. I. A. Gladkogo (Moscow: Nauka, 1965) p. 35.
27 Grossman, p. 45.
28 Granick, pp. 138–9.
29 Frank, p. 126.
30 Schapiro, p. 525n.
31 For example, over the period 1934–6 about two-thirds of glavk heads had engineering degrees and 60 per cent had been employed in the same branch of industry for more than ten years: see J. P. Hardt and T. Frankel, 'The Industrial Managers', in H. G. Skilling and F. Griffiths (eds), *Interest Groups in Soviet Politics* (Princeton University Press, 1971) p. 179.
32 McCagg, p. 307.
33 Hardt and Frankel, p. 179.
34 *Pravda* 27 December 1946, p. 1.
35 The phrase is McCagg's.
36 The Zhdanov-inspired purge of leading writers, painters and musicians from 1946 to 1948.
37 'KPSS v tsifrakh', *Partiinaya Zhizn'*, 1967, no. 19, p. 11.
38 Dunham, pp. 11–13 and B. Z. Zagitov, 'Nekotorye Voprosy Partiinogo Stroitel'stva v Pervye Poslevoemnye Gody (1945–50)', *Voprosy Partiinoi Raboty naSovremmenoi Etape* (Moscow: Mysl', 1969) p. 225.
39 Ibid., p. 226; however, 1,447,309 new candidates were admitted to the party between January 1946 and December 1951 (*Partiinaya Zhizn'*, 1973, no. 14, p. 12), indicating a very high rate of admissions in 1946 if the two sets of figures are really comparable.
40 This can be seen from a comparison of the figure for new admissions January 1947 to October 1952 of 580,000 with the actual increase in total membership of 655,638 in the period January 1947–January 1953. Rigby's estimate of an expulsion rate of over 100,000 p.a. in the years 1950–3 should thus be treated with caution; even if true, it implies an expulsion rate of only about 1.5 per cent p.a. See T. H. Rigby, *Communist Party Membership in the USSR 1917/67* (Princeton, University Press 1968) p. 281n. More accurate recent data suggest that some 540,000 left the CPSU for one reason or another (including death) in the period January 1952 to January 1956 (*Partiinaya Zhizn'*, 1973, no. 14).
41 For criticisms of party work at this time see *History of the Communist Party of the Soviet Union* (Moscow: 1960) p. 626–30, and, for example, resolutions attacking the work of Gorki obkom, and of three factory party committees (see *KPSS v Resolutsiyakh i Resheniyakh* (Moscow: Politizdat, 1971) tom. 6, pp. 277–80 and pp. 180–9).
42 L. Gruliow (ed.), *Current Soviet Policies* (New York: Praeger, 1953) pp. 56 and 82.
43 Ibid., pp. 59 and 81.
44 Ibid., p. 117.
45 See, W. J. Conyngham, *Industrial Management in the Soviet Union* (Stanford: Hoover, 1973) especially p. 47; and J. Berliner, *Factory and Manager in the USSR*

(Cambridge, Massachusetts: Harvard University Press, 1957) pp. 264–5.
46 A. P. Efimov, *Pravda*, 2 February 1951, p. 2.
47 Zagitov, p. 239.
48 Granick, p. 219, and Conyngham, p. 56.
49 Miller, p. 131.
50 See *PKh*, 1946, no. 5 (editorial) pp. 17–18.
51 See, for example, the speech by Kazakh First Secretary Zh. Shayakhmetov in *Pravda*, 25 February 1949.
52 Sh. Turetskii, in *PKh*, 1946, no. 4, p. 56.
53 Ibid., pp. 63–4; for a general description of such practices see Granick, pp. 145–7.
54 Granick, Chapter 10.
55 A. G. Zverev, *PKh*, 1947, no. 2, p. 33 (a billion = 1000 million).
56 V. Grossman, *PKh*, 1946, no. 3, p. 37.
57 B. Sukharevskii, *PKh*, 1947, no. 1, p. 21; A. V. Korobov, ibid., 1947, no. 3, p. 10.
58 Turetskii, p. 64.
59 See, for example, V. Grossman, *PKh*, 1946, no. 3, pp. 39–40.
60 Three such editorials in 1946 alone.
61 Berliner, p. 231.
62 See, *Izvestiya*, 5 and 6 July 1947, p. 1.
63 See, for example, *Current Digest of the Soviet Press*, vol. 1, 1949, no. 21, p. 54.
64 *Resheniya*, tom 3, pp. 362–8 and 603–14.
65 Arakelyan, p. 109.
66 Conyngham, p. 50.

3 The Command Economy and the Formation of Regional Policy 1945–53

1 L. M. Kantor, 'Perebazirovanie promyshlemosti SSSR', *Zapiski Leningradskogo Planovogo Instituta*, Vypusk VI, 1947, pp. 93–4.
2 The plan is reprinted in *Resheniya*, tom 2, pp. 676–703. The sections on regional policy are to be found on pp. 692–700.
3 D. M. Pinkhenson, *Ekonomicheskaya Geografiya v Tsifrakh* (Moscow: Prosveshchenie, 1970) pp. 36 and 38.
4 Ibid., p. 44.
5 This, it is argued, occurs only in the earlier stages of capitalist development (where Russia found herself in 1917). In the latter stages, as Lenin and others argued, markets in the developed regions became overcrowded and capitalists are forced to seek new markets in underdeveloped areas (like colonies) in order to survive.
6 These were Kiev, Khar'kov and Gorki in the Old West, Rostov-on-Don in the North Caucasus and also Sverdlovsk in the Urals—probably the most industrially developed city in the Near East at this time.
7 See D. Chernomordik, *VE*, 1948, no. 9, p. 17; and P. M. Alampiev, *Ekonomicheskoe Raionirovanie SSSR* (Moscow: Gosplanizdat, 1959) p. 185.
8 V. V. Kolotov, *Nikolai Alekseevich Voznesenskii* (Moscow: Politizdat, 1974) pp. 251–2.

9 I. Lerskii, *Vosproizvodstvo Osnovnykh Fondov Promyshlennosti SSSR v Usloviyakh Voiny* (Moscow: Gosplanizdat, 1945) pp. 26–7.
10 *Sbornik Soobschchenii Chrezvychainoi Gosudarstvennoi Kommissii o Zlodeyaniyakh Nemetsko-Fashistikh Zakhvatchikov* (Moscow: 1946) pp. 428–57.
11 N. A. Voznesenskii, *Voennaya Ekonomika SSSR v Period Otechestvennoi Voiny* (Moscow: 1948) pp. 56–7. It should be noted that not all enterprises 'put out of production' were *completely* destroyed; and A. V. Mitrofanova, *Rabochii Klass SSSR v Gody Velikoi Otechestvennoi Voiny* (Moscow: Nauka, 1971) p. 389. The figure probably only denotes All-Union enterprises and the higher figure of 96,500 (in *Shagi Pyatiletok*, (Moscow: Ekonomika, 1968, p. 127) all enterprises, including locally controlled ones.
12 *Eshelony Idut na Vostok; iz Istorii Perebazirovaniya Proizvodstvennykh Sil SSSR v 1941/1942 gg*. (Moscow: Nauka, 1966) (hereafter '*Eshelony*') pp. 108 and 140. The often-quoted figure of 1523 refers only to enterprises evacuated before December 1941; at least 150 enterprises were evacuated from behind the Leningrad and Stalingrad fronts in 1942–3.
13 V. B. Tel'pukhovskii, *Istoriya SSSR*, 1960, no. 6, p. 29.
14 *Istoriya Velikoi Otechestvennoi Voiny Sovetskogo Soyuza* (hereafter *IVOVSS*), tom II, pp. 147–8, and *Eshelony*, p. 108.
15 Lerskii, p. 17.
16 *IVOVSS*, tom II, p. 548.
17 B. Sokolov, *Promyshlennoe Stroitel'stvo v Gody Otechestvennoi Voiny* (Moscow: Gosplanizdat, 1946) p. 20.
18 Voznesenskii, pp. 109–10.
19 Ibid., p. 48, and *IVOVSS*, tom VI, p. 59.
20 *Sovetskaya Ekonomika v Period Otechestvennoi Voiny 1941/45 gg*. (Moscow: Nauka, 1970) (hereafter *SEOV*) pp. 157–8; *IVOVSS*, tom 2, p. 148; Lerskii, p. 22.
21 Soklov, pp. 20–1.
22 *IVOVSS*, tom II, p. 149, and Voznesenskii, p. 50.
23 G. S. Kravchenko, *Ekonomika SSSR v gody Velikoi Otechestvennoi Voiny* (Moscow: Ekonomika, 1970), pp. 352–3; 'means of destruction' are simply weapons and ammunition which have no reproductive impact on the economy.
24 Dokuchaev, p. 25.
25 Ibid., p. 20.
26 Ibid., pp. 11–14.
27 Voznesenskii, p. 60; *IVOVSS*, tom VI, p. 62; *SEOV*, p. 170; G. I. Shigalin, *Narodnoe Khozyaistvo v Period Otechestvennoi Voiny*, (Moscow: 1960) p. 54; *Resheniya*, tom 3, p. 49; Sokolov, p. 20.
28 Voznesenskii, p. 122; Shigalin, p. 145; *Resheniya*, tom 3, p. 37.
29 *IVOVSS*, tom VI, pp. 64–5.
30 *Resheniya*, tom 3, pp. 44–8.
31 Ibid., pp. 131–68. Much of this decree dealt with agriculture, food supplies and housing rather than industry and construction, but its publication coincided with the beginning of basic reconstruction (see Prikhod'ko, *Voss. Ind*., p. 73).
32 Kravchenko, p. 221 (figures at 1955 prices) and Sokolov, pp. 20–1.
33 Prikhod'ko, *Voss. Ind*., p. 76.
34 For example, see Sokolov, p. 32 and Kantor, p. 73.
35 *IVOVSS*, tom IV, p. 588.

36 Yu. A. Prikhod'ko, *Istoriya SSSR*, 1968, no. 6, p. 16.
37 *IVOVSS*, tom IV, p. 579; Prikhod'ko, *Voss. Ind.*, pp. 77 and 79.
38 Ibid., p. 116.
39 Ibid., pp. 53–5.
40 Both were sections of the Capital Construction Department, whose head, A. V. Korobov, later appeared as an ardent supporter of complex development.
41 Prikhod'ko, *Voss. Ind.*, p. 181.
42 G. G. Morekhina, *Voprosy Istorii*, 1961, no. 8, p. 59.
43 Prikhod'ko *Voss. Ind.*, p. 134.
44 Ibid.
45 *Narodnoe Khozyaistvo RSFSR* (Moscow: 1957) pp. 109–11; *Narodnoe Khozyaistvo SSSR v 1972 gg.* (Moscow: 1973) p. 168; Yu. A. Prikhod'ko, *Voss. Ind.*, pp. 181 and 217; Voznesenskii, pp. 49–51; G. A. Dokuchaev, *Rabochii Class Sibiri i Dalnego Vestoka v Gody Velikoi Otechestvennoi Voiny* (Moscow: Nauka, 1973) p. 39; data for the value of gross outputs of union republics are mostly based on locally published statistical handbooks of the period 1957–67 (mostly around 1962).
46 Ibid., p. 21; for data on transfer of labour see Ibid., p. 128 and *Trud v SSSR* (Moscow: 1967), pp. 42–71.
47 Some doubt has however been cast on the reliability of Soviet data on pre-war (and thus pre-Soviet) output levels in these areas.
48 Kantor, p. 96.
49 M. Z. Saburov appears to have held this post for a period during the war but Voznesenskii seems to have been restored to the post by 1945 (see Kolotov, p. 300).
50 The account that follows is based on that of Alampiev, pp. 185–7.
51 A. V. Korobov, *PKh*, 1941, no. 2, pp. 91–102.
52 E. Granovskii, *Bol'shevik*, 1945, no. 13, pp. 23–36.
53 See *PKh*, 1944, no. 3; and Lerskii.
54 See *PKh*, 1944, no. 2; and 1946 no. 1., pp. 25–36 and 67–79 respectively.
55 *Vedomosti Verkhovnogo Soveta*, 30 April 1945, 1, p. 48; and 2, p. 3.
56 Actually published in *Pravda* 9 February 1946, although it had been delivered some eight days earlier.
57 *Pravda*, 8 February 1946 (speech delivered 6 February 1946).
58 See *Bol'shevik*, 1945, no. 21, pp. 1–13 and pp. 14–19.
59 For example, see Jasny, pp. 248–9; Granick, Chapter 4.
60 For example, A. Zelenovskii, *PKh*, 1946, no. 1, p. 34; G. Sorokin, ibid., p. 63.
61 Many of the problems to be faced in the Five-Year Plan were also dealt with in projects for separate long-run plans for the liberated areas and for the Urals drawn up in 1943.
62 See *PKh*, 1946, nos 2 and 3, *passim*.
63 It may seem odd that Voznesenskii should have announced such a policy. The answer may well be that he was acting as an official spokesman for the leadership (as head of Gosplan), rather than voicing his personal opinions.
64 A. Arakelyan, *Osnovnye Zadachi Postevoennoi Pyatiletki* (Moscow: Gospolitizdat, 1946) pp. 45–6.
65 Voznesenskii, 'Speech on 5YP'. It should be noted that the investment data presented here refer to all investment in the economy, not just in industry.

However these figures do give a reasonable guide to the distribution of (planned) industrial growth as they also cover sectors outside industry but vital to future industrial growth, such as transport, agriculture, etc. The figures do exclude the major item in agricultural investment by the collective farms.

66 Sources: Zakon o pyatiletnem plan vosstanovleniya i razvitiya narodnogo khozyaistva SSSR na 1946/50 gg. (reprinted in *Resheniya*, tom 3, pp. 246–319; N. A. Voznesenskii, *Five Year Plan for the Rehabilitation and Development of the National Economy of the USSR 1946–50* (Moscow: 1946) (hereafter 'Speech on 5YP').
67 O. A. Konstantinov, *Voprosy Geografii*, sbornik 10, 1948, p. 66.
68 For a note on the calculation and significance of these ICORs see Appendix B.
69 Voznesenskii, 'Speech on 5YP', p. 9.
70 See *Resheniya*, tom 3, pp 332–4 and 430–1; bonuses etc. were to be paid to all workers signing agreements with the new Ministry of Labour Reserves, but for those agreeing to work in the east these allowances were approximately 50 per cent higher than for other regions.
71 *Izvestiya*, 1 March 1947, pp. 1–3; the shares in USSR total investment are lower than in the Five-Year Plan because of a much larger 'undistributed' item. Proportions amongst regions are basically similar.
72 A. Nove, *An Economic History of the USSR* (London: Penguin, 1969) p. 29.
73 *PKh*, 1946, no. 5, especially p. 21.
74 Ya. Feigin, *Bol'shevik*, 1946, no. 23–4, pp. 25–36.
75 Arakelyan, p. 46.
76 L. Volodarskii, *Vozrozhdenie Raionov SSSR postradavshchikh ot Nemetskoi Okkupatsii* (Moscow: Gosplanizdat, 1946) p. 12; and D. Degtyar', *Vozrozhdenie Rainov RSFSR podvergavshchikhsya Nemetskoi Okkupatsii*, (Moscow: Politizdat, 1947) p. 18.
77 *Izvestiya*, 2 August 1947, p. 3.
78 Ibid., 5 August 1947, p. 2.
79 Ibid., 12 August 1947, p. 3.
80 *VE*, 1948, no. 3, pp. 90–1.
81 Ibid., 1948, no. 8, p. 110.
82 M. Kozin, *Trud*, 24 May 1946, p. 2.
83 See N. I. Suprunenko, *Ukraina v Velikoi Otechestvennoi Voine Sovetskogo Soyuza (1941/1945 gg.)* (Kiev: 1956) pp. 400–1.
84 *Izvestiya* 15 January 1948, p. 2.
85 Ibid., 15 March 1949, p. 7.
86 Gruliow p. 140.
87 *Bol'shevik*, 1946, no. 15, p. 7.
88 I. Gladkov, *PKh*, 1947, no. 2, p. 84–9; and Ostrovityanov, *VE*, 1948, no. 1, pp. 86–92, and 1948, no. 8, pp. 66–75.
89 *VE*, 1948, no. 8, p. 83; and 1948, no. 9, p. 95.
90 Ibid., 1948, no. 8, p. 81.
91 *Uchenie Zapiski Leningradskogo Gosudarstvennogo Pedagogicheskogo Instituta im. A. I. Gertsena*, tom 47, kafedra Ekonomicheskoi Geografii i kafedra Metodiki Geografii (hereafter *LGPG*) pp. 125–48 (Nevel'shtein); 105–24 (Kolosovskii); 149–56 (Vol'f).

92 Yu. Saushkin, *Voprosy Geografii*, sbornik 6-oi, 1947, pp. 169–80.
93 *VE*, 1948, no. 8, pp. 101–10.
94 Saushkin, p. 173.
95 *VE*, 1948, no. 8, pp. 113–14.
96 For summaries of the debate see *Soviet Studies*, vol. I (1949–50) pp. 119–27 and vol. II (1950–1) pp. 317–22.
97 V. V. Novozhilov, in *Trudy Leningradskogo Finansovo-Ekonomicheskogo Instituta*, vypusk 3, 1947.
98 T. S. Khachaturov, *Osnovy Ekonomiki Zhelezdorozhnogo Transporta*, tom 1 (Moscow: 1946) and in *Izvestiya Akademii Nauk; otdel ekonomiki i prava* (hereafter *IAN(e)*) 1950, no. 4, pp. 252–68; P. Mstislavskii, *VE*, 1948, no. 7, pp. 131–5, and 1948, no. 10, pp. 35–48; S. G. Strumilin, *IAN(e)*, 1946, no. 3, pp. 195–215.
99 Vasyutin; E. S. Karnaukhova, *VE*, 1948, no. 9, pp. 61–3.
100 For example see *Soviet Studies*, vol. 1, p. 119, n. 2.
101 Khachaturov, *IAN(e)*, 1950, no. 4.
102 Khachaturov, *Osnovy* . . .
103 See *Resheniya*, tom 3, p. 736 (Fifth Five-Year Plan); A. G. Zverev, *Pravda*, 11 March 1949, pp. 2–5; Ya. E. Chadaev, *PKh*, 1950, no. 5, p. 84.
104 S. N. Prokopovich, *Narodnoe Khozyaistvo SSSR*, tom 1 (New York: Chekhov Publishing House, 1952) p. 15.
105 See B. Miroshnichenko, *PKh*, 1951, no. 3 (translated in *Soviet Studies* vol. 3, 1951–2, especially pp. 443–4).
106 *Resheniya*, tom 3, p. 717.
107 Gruliow (ed.), p. 152.
108 Ibid., p. 177.
109 Alampiev, p. 189.

4 The Implementation of Regional Policy After the War

1 Yu. I. Koldomasov, *Ekonomicheskii Svyazi v Narodnom Khozyaistve SSSR* (Moscow: Ekonomisheskaye Literatura, 1963) p. 17.
2 As Table 3.2 and *Narodnoe Khozyaistvo RSFSR* (Moscow: 1957) pp. 109–11.
3 I.e. the Baltic States, right-bank Moldavia and the westernmost oblasti of the Ukraine and Belorussia. The data are derived from Prikhod'ko, *Voss. Ind.*, p. 231.
4 *Narodnoe Khozyaistvo RSFSR*, pp. 112–14; *Narodnoe Khozyaistvo SSSR v 1958 godu* (Moscow: 1959) p. 141.
5 The data here refer to the post-1961 Urals economic region, which differs from the pre-1961 version in incorporating the Tyumen and Kurgan oblasti and excluding the Bashkir ASSR.
6 As Table 3.2 and *Kapital'noe Stoitel'stvo v SSSR* (Moscow: 1961) pp. 110, 111, 117; Dokuchaev, *1946–50*, p. 26.
7 Dokuchaev, *1946–50*, p. 35.
8 For example, M. Sonin, *VE*, 1948, no. 6, p. 20, and the decrees on incentives for workers going to the Eastern regions (n. 70 to Chapter 3).
9 *Trud v SSSR*, pp. 42–71; Dokuchaev, *1941–5*, p. 43; Dokuchaev, *1946–50*, p. 48.

10 *Rabochie Sibiri v Bor'be za Postroenie Sotsializma i Kommunizma (1917/67, gg.)* (Kemerovo: 1967) p. 247.
11 Dokuchaev, *1946–50*, p. 66.
12 *Istoriya Rabochego Klassa Leningrada*: vypusk 1 (Leningradskii Universitet, 1962) pp. 113–4; *Pravda*, 3 April 1946, p. 2; *Izvestiya*, 13 August 1947, p. 2.
13 Dokuchaev, *1946–50*, p. 63.
14 *Resheniya*, tom 3, pp. 342–3.
15 Dokuchaev, *1946–50*, pp. 79–80 and 86.
16 *Pravda*, 2 October 1952.
17 See Prikhod'ko, *Voss. Ind.*, p. 191; and Dokuchaev, *1946–50*, p. 82.
18 For example, see Prikhod'ko, *Voss. Ind.*, pp. 188 and 198.
19 Dokuchaev, *1946–50*, pp. 66–7.
20 Gladkov, p. 35.
21 Dokuchaev, *1946–50*, p. 65.
22 G. Kosyachenko, *PKh*, 1946, no. 2, p. 144.
23 *Izvestiya*, 30 July 1947, p. 2.
24 *VE*, 1948, no. 4, p. 82, and *PKh*, 1946, no. 6, p. 8.
25 *Izvestiya*, 8 January 1948, p. 3.
26 See A. Chernyak, *O Tempakh Razvitiya Sotsialisticheskoi Promyshlennosti* (Moscow: Gospolitizdat, 1948) p. 61; and *KPSS v Rezolyutsiyakh i Resheniyakh*, tom VI, p. 202.
27 *Izvestiya*, 15 July 1947, pp. 1 and 3; and 8 January 1948, p. 2.
28 Ibid., 19 July 1947, p. 1.
29 *PKh*, 1950, no. 5, p. 82.
30 Gruliow, p. 112.
31 Dokuchaev, *1946/50*, pp. 56–7.
32 See Prikhod'ko, *Voss. Ind.*, p. 226.
33 See the decree of 28 December 1951 in *Resheniya*, tom 3, pp. 674–7.
34 See sources to Tables 3.1, 4.1 and 4.4.
35 *Izvestiya*, 30 August 1947, p. 2.
36 Ibid., 29 August 1947, p. 2.
37 *PKh*, 1947, no. 2, p. 9.
38 *Izvestiya*, 30 July 1947, p. 2.
39 Ibid., 8 January 1948, p. 2.
40 See *Resheniya*, tom 3, pp. 603–14.
41 *Kazakhstanskaya Pravda*, 21 September 1952, p. 3.
42 *Pravda*, 12 October 1952, p. 1.
43 *PKh*, 1947, no. 3, p. 5.
44 *Pravda*, 16 May 1946, p. 2; and 18 May 1946, p. 2.
45 Gruliow, p. 174.
46 Dokuchaev, *1946–50*, p. 24.
47 *Izvestiya*, 9 July 1947, p. 2.
48 *PKh*, 1950, no. 5, especially pp. 84–6.
49 *Resheniya*, tom 3, p. 611.
50 *Izvestiya*, 10 July 1947, p. 2.
51 Ibid., 18 September 1947, p. 2.
52 *Rabochie Sibiri . . .* (see n. 10) p. 247; *Izvestiya* 7 August 1947, p. 1 (editorial).
53 D. Fish, *Voprosy Truda i Zarabotnoi Platy v sotsialisticheskoi Promyshlennosti* (Leningrad: Lenizdat, 1948) p. 19; and *VE*, 1948, no. 6, p. 122.

54 *Resheniya*, tom 3, p. 468.
55 See L. Maizenberg, *PKh*, 1950, no. 6. pp. 58–70.
56 Gruliow, p. 112.
57 Based on Koldomasov, pp. 30–4.
58 *Izvestiya*, 29 August 1947, p. 2; and Gruliow, p. 117.
59 A. Galitskii, *PKh*, 1946, no. 2, pp. 115–16.
60 L. Sapozhnikov and L. Ulitskii, *PKh*, 1947, no. 4, p. 40.
61 *VE*, 1948, no. 4, p. 2.
62 For example, see ibid.; *Pravda* 18 May 1946, p. 1 (editorial); Galitskii, p. 119.
63 Sh. Turetskii, *PKh*, 1946, no. 4, p. 66.
64 D. Chernomordyak, *VE*, 1948, no. 9, p. 17.
65 Maizenberg, p. 60.
66 Gruliow, p. 130.
67 Dokuchaev, 1946–50, pp. 112–13.
68 Examples of criticism of '*vedomstvennost'*' (departmentalism) may be found in *PKh*, 1946, no. 4, p. 15; and 1947, no. 2, p. 35; also in Malenkov's speech to the 19th Party Congress.
69 Rubin, p. 143.
70 Gladkov, p. 28.
71 N. N. Kolosovskii, in *Voprosy Geografii*; sbornik VI, p. 135.
72 A. Arakelyan, *Osnovnye Zadachi Poslevoennoi Pyatiletki* (Moscow: Gospolitizdat, 1946) p. 49.
73 A. Arakelyan, *Industrial Management in the USSR*, pp. 139–41.
74 *PKh*, 1950, no. 5, p. 86.
75 For example see *Voprosy Geografii*, sbornik VI, *passim*.
76 See *Partiinaya Zhizn'* 1948, no. 5, pp. 6–19; and Gruliow, pp. 76 and 81.
77 Gruliow, pp. 118 and 135–7.
78 For example, see *PKh*, 1946, no. 1 p. 36; 1946, no. 4, p. 17; and 1946, no. 5, p. 25.
79 Kolotov, p. 251.
80 *Kazakhstanskaya Pravda*, 24 February 1949.
81 See, for example, S. Swianiewicz, *Forced Labour & Economic Development*, (London: Oxford University Press, 1965).
82 Rovinskii, p. 46.
83 But see the argument presented below in Chapter 6.
84 *PKh*, 1950, no. 5, pp. 82–90.
85 Much of Siberian local and co-operative industry was devoted to the production of light industrial goods, the total output of which in Siberia and the Far East fell slightly over the war years: see Dokuchaev, *1946–50*, p. 20.
86 *PKh*, 1950, no. 5, pp. 84–5.
87 *Pravda*, 14 December 1946, p. 1.
88 *PKh*, 1950, no. 5, pp. 65 and 82.
89 R. S. Livshits, *Ocherki po Razmeshcheniyu Promyshlennosti SSSR* (Moscow: Gospolitizdat, 1954) p. 34.
90 See *Resheniya*, tom 3, pp. 519–30 and 652–4.
91 For example see *Sovetskaya Sotsialisticheskaya Ekonomika 1917/1957 gg*. (Moscow: 1957) p. 491.
92 Dokuchaev, 1946–50, p. 45 (Table 7).

93 For example, see A. Nove, 'The Industrial Planning System: The Reforms in Retrospect', *Soviet Studies*, vol. 14, 1962–3, especially pp. 3–4.

5 The Formation of Sectoral Policy 1945–53

1 In practice the 'A' and 'B' sectors are not completely synonymous with 'heavy' and 'light' industry; some of 'A' sector output is light industrial and of 'B' sector heavy. The best indicator of the heavy/light balance in Soviet statistics is, however, the A/B balance.
2 *Narodnoe Khozyaistvo SSSR v 1972 g*, p. 162.
3 Nove, *Economic History* . . . , pp. 78–82.
4 J. V. Stalin, 'The Tasks of Business Executives' (speech of 4 February 1931), in J. V. Stalin, *Leninism* (London: Lawrence & Wishart, 1940) p. 366.
5 This formulation is suggested by, for example the input–output analysis of political systems provided by Easton, Almond and others.
6 V. Dyachenko, *VE*, 1948, no. 1, p. 43.
7 *Narodnoe Khozyaistvo v 1972 g*., p. 162; Kravchenko, p. 351.
N.B. The first source gives a figure of 25.1 per cent for the 'B' sector in 1945.
8 *IVOVSS*, tom VI, p. 78.
9 *SEOV*, p. 118.
10 Prikhod'ko, *Voss. Ind*. It should be noted that the term 'light industry' is here used to denote the entire 'B' sector; the non-foodstuffs sector is termed 'pure light industry'. The same term is often and confusingly used for both.
11 Ibid., p. 21.
12 *IVOVSS*, tom II, p. 149.
13 *SEOV*, p. 121.
14 *IVOVSS*, tom VI, p. 64.
15 Kantor, p. 77.
16 Prikhod'ko, *Voss. Ind*. pp. 106–7 and 116; *SEOV*, p. 118.
17 'A' sector output (including the armaments industries) was 36 per cent above its 1940 level in 1944, but fell to only 82 per cent of that level in 1946, as the process of reconversion to civilian production was completed. It regained its pre-war level in 1947 however. 'B' sector output regained its 1940 level only in 1949.
18 I. Kuzminov, *Bol'shevik*, 1945, nos. 17–18, p. 35.
19 A. Zelenovskii, *PKh*, 1946, no. 1, pp. 33–4.
20 A. Korobov, *PKh*, 1946, no. 3, p. 20.
21 A. G. Zverev, *Zapiski Ministra* (Moscow: Politizdat, 1973) pp. 227–30. During the war Soviet citizens accumulated cash as there was very little upon which to spend it. This surplus of spending power threatened severe inflationary pressures after the war unless the supply of consumer goods was greatly expanded. Indeed a flourishing black market existed during our period and these 'surplus' roubles were only partially absorbed by the currency reform of 1947 (which cut the value of savings in banks by ten) and the establishment of higher 'commercial' prices for some categories of goods.
22 See, for example, Conquest, *Power and Policy* . . . Chapter 4.
23 *Bol'shevik*, 1945, no. 21, p. 12.
24 Ibid., p. 10.

25 Ibid., p. 14.
26 *Pravda*, 10 February 1946, p. 1.
27 See nn. 28 and 29 below.
28 *PKh*, 1946, no. 5, p. 7.
29 McCagg, pp. 134–5.
30 *Resheniya*, tom 3, pp. 250–2.
31 See Voznesenkii, 'Speech on 5YP', p. 13.
32 The way in which Voznesenskii's announced targets fit in with more detailed sectoral targets is well illustrated in A. Bergson, 'The Fourth Five Year Plan: Heavy versus Consumers' Goods Industries', *Political Studies Quarterly*, vol. LXII, no. 2, pp. 220–2.
33 *Narodnoe Khozyaistvo v 1972 g*., p. 162; Voznesenskii, 'Speech on 5YP' p. 13.
34 A rough calculation suggests that the output of the armaments industry fell by as much as 70 per cent in 1945–6; the output of civilian industry rose by 20 per cent, whilst that of industry as a whole fell by 16 per cent. (Data on civilian industry from *PKh*, 1947, no. 1, p. 25.)
35 *Resheniya*, tom 3, pp. 252–72; *The USSR Economy: A Statistical Abstract* (London: Lawrence & Wishart, 1957).
36 Arakelyan, *Osnovnye Zadachi* . . . p. 38.
37 B. Braginskii and A. Vikentev, *VE*, 1948, no. 3; p. 18.
38 The plan called on local and co-operative industry to specialise in the production of mass consumption goods (*Resheniya*, tom 3, p. 271).
39 Ibid., p. 364; and *Kapital'noe Stroitel'stvo v SSSR*, pp. 66–7.
40 *Pravda*, 28 April 1946, p. 2 (interview with V. P. Zotov).
41 See Voznesenkii, 'Speech on 5YP', p. 13.
42 *Pravda*, 2 March 1946, p. 3; the problems were mainly concerned with disruption of work schedules and over-diversification of effort during reconversion.
43 Ibid., 30 May 1946, p. 1.
44 *Resheniya*, tom 3, pp. 350–62 and 363–8.
45 *Pravda*, 29 December 1946, p. 1.
46 For example, see McCagg, p. 293.
47 Ibid.
48 Gladkov, p. 8; Prikhod'ko, *Voss. Ind*., p. 238.
49 Vikentev, p. 70.
50 L. Gatovskii, *Ekonomicheskaya Pobeda Sovetskogo Soyuza v Velikoi Otechestvennoi Voine* (Moscow: Politizdat, 1946) p. 119.
51 Arakelyan, p. 19.
52 E. Lokshin, *PKh*, 1946, no. 4, p. 43.
53 G. P. Kosyachenko, *PKh*, 1946, no. 4, p. 9.
54 *PKh*, 1946, no. 5, pp. 16–17.
55 V. Shatalov, *PKh*, 1947; no. 3, pp. 15–25.
56 For example, *Pravda*, 30 May 1946, p. 1 (editorial); 29 December 1946, p. 1 (editorial); and 11 January 1949, p. 2.
57 Ibid., 12 October 1952, p. 5.
58 See *VE*, 1950, no. 10, pp. 99–108 for a summary of part of the debate.
59 For example, *Pravda*, 13 April 1946, p. 1; Kosyachenko, *PKh*, 1946, no. 4, p. 4.
60 A. Petrov, *PKh*, 1947, no. 2, pp. 66 and 58.
61 *VE*, 1948, no. 1, pp. 91–2.

62 The following account is based on that of Anchishkin (see n. 58).
63 I. Stalin, 'Ob Oshibkakh t Yaroshenko L. D.', in *Ekonomicheskie Problemy Sotsializma v SSSR* (Moscow: Gospolitizdat, 1952) pp. 58–83.
64 Ibid., pp. 66–7.
65 See A. Katz, *The Politics of Economic Reform in the Soviet Union*, (New York: Praeger, 1972) pp. 38–9.
66 Ibid., p. 38.
67 M. A. Suslov, *Pravda*, 24 December 1952, p. 2. Suslov attacked Fedoseev and *Bol'shevik* for unduly praising Voznesenskii's book and for favouring the 'un-Marxist' point of view that there was no immutable law of the development of socialist economies. Voznesenskii and others had argued that that development was merely the result of the government's and the party's economic policy.
68 L. I. Skvortsov, *Tseny i Tsenobrazovanie v SSSR* (Moscow: Vyssnaya Shkola, 1972) pp. 84–7.
69 See Conquest, *Power and Policy*, pt. II *passim*.
70 Stalin, *Ekonomicheskie Problemy* . . . pp. 55–6.
71 Vikentev, pp. 72–3.
72 See Katz, p. 45.

6 The Implementation of Sectoral Policy after the War

1 *Narodnoe Khozyaistve v 1972 g.*, *passim*; *Resheniya*, tom 3, pp 252–72; *The USSR Economy* (London: 1957) *passim*; *Izvestiya*, 17 April 1951, pp. 1–2.
2 For example, Gladkov, p. 7; Lokshin, *Promyshlennost SSSR*, pp. 119–20; Rubin, p. 136.
3 A. Arakelyan, *Vedushaya Rol' Promyshlennosti v Razvitti Narodnogo Khozyaistva SSSR* (Moscow: *Pravda*, 1951) p. 10.
4 No doubt the practice of overweighting some types of output by using 1926–7 prices to aggregate production data from different branches was partly to blame for this (see Nove, *Economic History*, pp. 381–6).
5 *Izvestiya*, 17 April 1951, p. 1.
6 See for example, *Pravda*, 18 February 1949, p. 1; and 19 February 1949, p. 1.
7 Calculated from *Nar. Khoz. 1972*, p. 195.
8 *Resheniya*, tom 3, p. 363; *The USSR Economy*.
9 Prikhod'ko, *Voss. Ind.*, p. 212.
10 Calculated from *Nar. Khoz. 1972*, pp. 159–62; *Kap. Stroi*, pp. 66–7.
11 Calculated from *Nar. Khoz. 1972*, p. 160.
12 *Kap. Stroi*, pp. 66–7; and *Resheniya*, tom 3.
13 These two ministries were amalgamated into a new Ministry of Light Industry in December 1948.
14 The dramatic fall in the machine-building sector's share may be due to the reorganising of ministries during this period. The wartime Ministries of Heavy and Medium Machine-Building and of Machine Tool Manufacture saw many of their enterprises 'hived off' in 1945–6 to four new ministries, at least three of which specialised in servicing individual sectors of the economy (transport, agriculture and construction). The investments of these ministries

(and of the Automobile Industry Ministry, based on the former Ministry of Medium Machine-Building) may have been included in other categories in post-war statistics. The machine-building sector could otherwise surely have not comfortably overfulfilled its output plan in 1950.

15 *PKh*, 1946, no. 6, p. 5.
16 Ibid., 1947, no. 2, p. 8.
17 A. Zverev, *PKh*, 1947, no. 2, p. 33.
18 *Izvestiya*, 19 March 1949, p. 1 (editorial).
19 See Prikhod'ko, *Voss. Ind.*, p. 237.
20 V. Dyachenko, *VE*, 1948, no. 1, p. 43.
21 V. Grossman, *PKh*, 1946, no. 3, pp. 36–47, *passim*.
22 N. Eremenko, *PKh*, 1946, no. 2, p. 93; A. Korobov, *PKh*, 1946, no. 3, p. 33.
23 Prikhod'ko, *Voss. Ind.*, p. 33.
24 A. Korobov, *PKh*, 1947, no. 3, pp. 11–12.
25 *Izvestiya*, 17 January 1948, p. 2.
26 Prikhod'ko, *Voss. Ind.*, p. 212.
27 *Resheniya*, tom 3, p. 362.
28 Ibid.; *Pravda*, 30 May 1946, p. 1 (editorial) and 29 December 1946, p. 1 (editorial).
29 *Trud v SSSR*, p. 42. The figure referred to is 'number of manual workers and employees'.
30 Mitrofanova, pp. 442–3.
31 Prikhod'ko, *Voss. Ind.*, p. 191.
32 V. P. Zotov, *Razvitie Pishchevoi Promy shlennosti v Novoi Pyatiletki* (Moscow: Gospolitizdat, 1947) p. 36.
33 Zverev, *Zapiski Ministra*, p. 227.
34 *Pravda*, 20 February 1946, p. 1 (editorial).
35 *Trud v SSSR*, pp. 84–5.
36 *Resheniya*, tom 3, p. 343.
37 Ibid., p. 428.
38 Ibid., pp. 674–7.
39 Ibid., p. 284.
40 I. M. Brover, *Ocherki Razvitiya Tyazheloi Promyshlennosti SSSR* (Alma Ata: AN Kaz. SSR, 1954), p. 238.
41 *Pravda*, 27 December 1946, p. 2.
42 For example see *PKh*, 1946, no. 1, p. 98; and 1946, no. 2, p. 144.
43 Shatalov, *PKh*, 1947, no. 3, p. 23.
44 Lokshin, p. 147; Brover, pp. 231, 232 and 239.
45 Prikhod'ko, *Voss. Ind.*, p. 212.
46 Vikentev, pp. 98 and 154.
47 *PKh*, 1946, no. 6, p. 9.
48 Shatalov, pp. 21–2.
49 See above, Chapter 3; *PKh*, no. 4, p. 13.
50 At 1940 = 100, output per unit of machinery per hour in 1950 was as follows: Cotton spindles – 100, Cotton looms – 95, Wool spindles – 104, Wool looms – 104 (calculated from data in *The USSR Economy*, pp. 85–7).
51 Lokshin, p. 138.
52 *Izvestiya*, 30 August 1947, p. 2.
53 The drought of 1946 lead to a famine in some areas, and the (low) 1936–40

average grain harvest levels were regained only in 1950 and exceeded by only 9 per cent over the period 1951–3.

54 See Kosygin's speech to the 19th Congress (*Pravda*, 13 October 1952, p. 6).
55 Lokshin, pp. 119–20.
56 Shatalov, p. 22.
57 *Pravda*, 18 February 1949, p. 2.
58 *Resheniya*, tom 3, p. 362.
59 Zverev, p. 227.
60 Calculated from *Prom. v SSSR*, pp. 104, 138, 139, 190, 206, 242, 243, 275, 321, 369.
61 *Izvestiya*, 13 September 1947, p. 2.
62 Speech by L. P. Lykova, Printed in *Pravda*, 7 October 1952, p. 5.
63 H. Hunter, *Soviet Transportation Policy* (Washington DC: Brookings Institution, 1968) p. 324.
64 Prikhod'ko, *Voss. Ind.*, p. 237.
65 *Resheniya*, tom 3, p. 366.
66 Prikhod'ko, *Voss. Ind.*, pp. 169–70.
67 *Resheniya*, tom 3, pp. 603–14.
68 Prikhod'ko, *Voss. Ind.*, p. 215.
69 Skilling and Griffiths, pp. 171–87.
70 *Pravda*, 2 March, 1946, p. 3.
71 For example, see *Pravda*, 29 December 1946, p. 1.
72 *Pravda*, 1 January 1949, p. 2.
73 It has been argued that to increase 'B' sector output one needs machines, construction materials and other supplies that are the product of 'A' sector industries. Therefore to increase 'B' sector output one must first expand 'A' sector ouput. Brover argued in 1954 that this was in fact done during the Fourth Five-Year Plan period (Brover, p. 265). He asserted that the basis for rapid expansion of the 'B' sector during the Fifth Plan period had been established by great increases in the output of machinery and materials by the 'A' sector for the 'B' sector over the period 1946–50. The data on supplies to light industry over that period presented above, however, must cast doubt on his assertions. Supplies were certainly far below their planned level. The bulk of expansion of 'A' sector output was taken up by supplies to industries within that sector. That was surely one of the main reasons why the Fifth Five-Year plan period saw the 'A' sector grow faster than the 'B' sector (90 per cent compared to 76 per cent) in spite of Malenkov's initiative on consumer goods and Khrushchev's on foodstuffs.
74 *Pravda*, 2 March 1946, p. 3.
75 Ibid., 27 December 1946, p. 1 (editorial).
76 See Schapiro, *CPSU*, pp. 509–11.
77 See Gruliow, p. 117.
78 *Pravda*, 2 January 1951, p. 2.
79 Zelenovskii, *PKh*, 1946, no. 1, p. 35; Kosyachenko, *PKh*, 1946, no. 4, p. 8.
80 Ibid., (Kosyachenko), p. 11.
81 Skvortsov, p. 84; an even higher figure (41.2 billion roubles) is given by A. Malafeev, *Istoriya Tsenoobrazovaniya v SSSR* (Moscow: 1964) pp. 246 and 252.

82 See L. Maizenberg, *PKh*, 1950, no. 6, p. 66; and Sh. Turetskii, ibid., 1946, no. 4, p. 64.
83 Kosyachenko, ibid., 1946, no. 4, p. 15; Gruliow, p. 118.
84 V. Grossman, ibid., 1946, no. 3, p. 39.
85 *Pravda*, 20 November 1946, p. 1 (editorial).
86 Ibid., 26 October 1949, p. 3.
87 Zverev, *PKh*, 1947, no. 2, p. 34.

7 Conclusion

1 See, for example, McCagg, Conquest, Berliner, and Granick.
2 Medvedev, p. 490; Werth, p. 283.
3 See McCagg, p. 307 and Conquest, Chapter 8.
4 A. G. Meyer, 'USSR Inc.', *Slavic Review*, 1961, pp. 369–76.
5 See above, p. 15.
6 Quoted in R. E. Neustadt, *Presidential Power* (New York: Wiley, 1960) p. 9.
7 See, for example, Katz, *passim*.

Appendixes

1 See N. Jasny, *Soviet Statistical Handbook*; G. W. Nutter, *The Growth of Industrial Production in the Soviet Union* (Princeton University Press, 1962); D. R. Hodgman, *Soviet Industrial Production 1928–51* Cambridge Massachusetts; (Harvard University Press, 1954). For a summary of the problems of assessing Soviet growth rates see Nove, pp. 381–6.
2 This argument is spelt out more fully in A. Bergson, 'The Fourth Five-Year Plan: Heavy versus Consumer Goods Industries', *Political Science Quarterly*, LXII, no. 2 220–7.

Bibliography

Western Studies of Soviet Politics 1945–53

Conquest, R., *Power & Policy in the USSR* (London: Macmillan, 1961).

Deutscher, I., *Heretics and Renegades* (London: Jonathan Cape 1955).

Deutscher, I., *Stalin: a political biography* (Harmondsworth, Penguin, revised edn, 1966) Chapter 15.

Djilas, M., *Conversations with Stalin* (Harmondsworth: Pelican, 1969).

Dunham, V. S., *In Stalin's Time: Middle Class Values in Soviet Fiction* (Cambridge, 1976).

Frankel, E. R., 'Literary Policy under Stalin in Retrospect: A Case Study 1952–3', in Shapiro, J. P., and Potichnyj, P. T. (eds), *Change and Adaptation in Soviet and East European Politics* (New York: Praeger, 1976).

Inkeles, A., *Social Change in Soviet Russia* (Cambridge, Massachusetts: Harvard University Press, 1968) Chapter 15.

Khrushchev Remembers, vol. 1 (London: Sphere Books, 1971); includes text of the 'Secret Speech' of 1956).

McCagg, W. O., Jr, *Stalin Embattled 1943/8* (Detroit: Wayne State University Press, 1978).

Medvedev, R., *Let History Judge* (London: Spokesman Books, 1976).

Moore, B., *Terror and Progress USSR* (Cambridge, Massachusetts: Harvard University Press, 1954).

Nemzer, L., The Kremlin's Professional Staff: the 'Apparatus' of the CC, CPSU, *American Political Science Review*, 1950, pp. 64–85.

Pethybridge, R., *A History of Post-War Russia* (London: Allen & Unwin, 1966).

Ploss, S., *Conflict & Decision making in Soviet Russia: A Case Study of Agricultural Policy 1953–63* (Princeton University Press 1969) pp. 1–58.

Rigby, T. H., *Communist Party Membership in the USSR 1917–67* (Princeton University Press, 1968).

Shapiro, L., *The Communist Party of the Soviet Union* (London: Methuen, 1963).

Shulman, M., *Stalin's Foreign Policy Reappraised* (Cambridge, Massachusetts: Harvard University Press, 1963).

Werth, A., *Russia at War 1941–5* (London: Barrie & Rockliff, 1964).
Werth, A., *Russia: The Post-War Years* (London: Hale, 1971).
Wolfe, B. D., *An Ideology in Power* (London: Allen & Unwin, 1969).

2 Western Studies of the Command Economy

Berliner, J., *Factory and Manager in the USSR* (Cambridge, Massachusetts: Harvard University Press, 1957).
Connolly, V., *Beyond the Urals: Economic Developments in Central Asia* (Oxford University Press, 1967).
Conyngham, W. J., *Industrial Management in the Soviet Union: the CPSU in Industrial Decision–Making 1917–1970* (Stanford: Hoover, 1973).
Fakiolas, R., 'Problems of Labour Mobility in the USSR', *Soviet Studies*, 1962–3, pp. 16–35.
Frank, A. G., 'The Organisation of Economic Activity in the Soviet Union', *Weltwirtschaftliches Archiv*, 1957, pp. 104–56.
Granick, D., *Management of the Industrial Firm in the USSR*, (Columbia University Press, 1954).
Hardt, J. P., and Frankel, T., 'The Soviet Industrial Managers', in Skilling, H. G., and Griffiths, F., *Interest Groups in Soviet Politics* (Princeton University Press, 1971).
Hunter, H., *Soviet Transport Experience: Its Lessons for Other Countries* (Washington DC: Brookings Institution, 1968).
Hutchings, R., Geographic Influences on Centralisation in the Soviet Economy', *Soviet Studies*, 1965–6, pp. 286–302.
Inch, P. J., 'The Evolutionary Process of Soviet Regional Economic Planning', (unpublished M. Phil. thesis, University of London, 1970).
Jasny, N., *The Soviet 1956 Statistical Handbook: A Commentary* (East Lancing: Michigan State University Press, 1957).
Jasny, N., *Soviet Industrialisation 1928–52* (Chicago University Press., 1961).
Koropeckyj, I. S., 'The Development of Soviet Location Theory before the Second World War', *Soviet Studies*, 1967–8, pp. 1–28 and 232–44.
Mieczkowski, Z., 'The Economic Regionalization of the Soviet Union in the Lenin and Stalin period', *Canadian Slavonic Papers VIII*, 1966, pp. 89–124.
Miller, J., 'Soviet Planners in 1936/7, in Degras, J., and Nove, A (eds), *Soviet Planning: Essays in Honour of Naum Jasny* (Oxford: Blackwell, 1964).
Prokopovich, S. N., *Narodnoe Khozyaistvo SSSR* (New York: 1952).
Nove, A., *An Economic History of the USSR* (Harmondsworth: Penguin, 1969) Chapters 10 and 11.

Vucinich, A., *Soviet Economic Institutions* (Stanford University Press, 1952).

Articles and Books Published in the USSR Since 1953

Alampiev, P. M., *Ekonomicheskoe Raionirovaniye SSSR* (Moscow: Gosplanizdat, 1959).

Aleksandrov, A., *Trud Vo Imya Pobedy* (Syktyvkar: Komi Knizhnoe Izd–Vo, 1968).

Brover, I. M., *Ocherki Razvitiya Tyazheloi Promyshlennosti SSSR* (Alma Ata: AN Kaz SSR, 1954).

Chadaev, Ya. E., *Ekonomika SSSR 1941–1945 gg* (Moscow: 1965).

Deborin, G. A., and Tel'pukhovskii, B. S., *Itogi I Uroki Velikoi Otechestvennoi Voiny* (Moscow: Mysl', 1970).

Dokuchaev, G. A., *Rabochii Klass Sibiri I Dal'nego Vostoka v Poslevoennve Gody* (Novosibirsk: Nauka, 1972).

Dokuchaev, G. A., *Rabochii Klass Sibiri i Dal'nego Vostoka v Gody Velikoi Otechestvennoi Voiny* (Moscow: Nauka, 1973).

Dvoinishnikov, M. A., *Deyatel'nost' KPSS po Vosstanovlenniyu Promyshlennosti v Poslevoennoi Period* (Moscow: Politizdat, 1977).

Efimov, A. N. (ed.), *Ekonomika SSSR v Poslevoennoi Period* (Moscow, 1967).

Ekonomicheskaya Istoriya SSSR, Pod. red. I. S. Golubnichogo *et al* (Moscow: Mysl', 1967).

Entelis, G. S., in *Izvestiya AN Moldavskoi SSSR: Seriya Obshestvennykh Nauk*, 1971, no. 2, pp. 3–10.

Eshelony Idyt Na Vostok: Iz Istorii Perebazirovaniya Proizvodstvennvjkh Sil SSSR v 1941–1942 gg (Moscow: Nauka, 1968).

Formirovanie i Razvitie Sovetskogo Rabochego Klassa (1917–1961 gg): *Sbornik Statei* (Moscow: Nauka, 1964).

Gvozdev, B. I., in *Vestnik MGU: Istoriya*, 1971, no. 5, pp. 3–15.

History of the Communist Party of the Soviet Union (Moscow: 1960).

Istoriya Rabochego Klassa Leningrada: vypusk 1 (Leningradskii Universitet, 1962).

Istoriya Velikoi Otechestvennoi Voiny Sovetskogo Soyuza 1941–45 gg tom 2, 4, 6 (Moscow: Voenizdat, 1961, 1963, 1965).

Koldomasov, Yu. I, *Ekonomicheskie Svyazi v Narodnom Khozyaistve SSSR* (Moscow: Ekonomicheskaya Literatura, 1963) p. 17.

Kukin, D. M., in *Voprosy Istorii 1971*, no. 8, pp. 27–42.

Kolotov, V. V., *Nikolai Alekseevich Voznesenskii* (Moscow: Politizdat, 1974).

Kolotov, V. V. in *Voprosy Istorii KPSS*, 1963, no. 6, pp. 94–8.

Kravchenko, G. S., *Ekonomika SSSR v gody Velikoi Otechestvennoi Voiny* (Moscow: Ekonomika. 1970).

Livshits, R. S. *Ocherki po Razmeshcheniyu Promyshlennosti SSSR* (Moscow: Gospolitizdat, 1954).
Lokshin, E. Yu., *Promylennost' SSSR 1940–63* (Moscow: Mysl', 1964).
Na Novgorodskom Zemle: Sbornik Statei (Novgorod: 1968).
Morekhina, G. G., *Voprosy Istorii*, 1961, no. 8, pp. 41–60.
Ocherki Istorii Velikoi Otechestvennoi Voiny 1941–1945 gg (Moscow: AN SSSR, 1955).
Ocherki po Istorii Narodnogo Khozyaistva SSSR: Sbornik Statei (Moscow: Politizdat, 1959).
Popov, V. E., Shapiro, I. S., *Chernaya Metallurgiya Sibiri* (Moscow: AN SSSR, 1960).
Plenum Tsentral' nogo Komiteta Kommunisticheskoi Partii Sovetskogo Soyuza, 18–21 Iyunya 1963 g.: Stenograficheskii Otchet (Moscow: Politizdat, 1964).
Prikhod'ko, Yu. A., *Vosstanovlenie Industrii 1942–50* (Moscow: Mysl', 1973).
Prikhod'ko, Yu. A., in *Istoriya SSSR*, 1968, no. 6, pp. 9–24.
Prikhod'ko, Yu. A., in *Voprosy Istorii*, 1969, no. 5, pp. 25–36.
Rabochie Sibiri v Bor'be za Postroenie Sotsializma i Kommunizma 1917–1967 gg. (Kemerovo, 1967).
Razvitiye Sotsialisticheskoi Ekonomiki SSSR v Poslevoennoi Period, pod. red. I. A. Gladkova (Moscow: Nauka, 1965).
Rubin, A. M., *Organizatsiya Upravleniya Promyshlennostyu v SSSR (1917–1967 gg)* (Moscow: Ekonomika, 1969).
Shagy Pyatiletok: Razvitie Ekonomiki SSSR (Moscow: Ekonomika, 1968).
Shigalin, G. I., *Narodnoe Khozyaistvo SSSR v Period Velikoi Otechestvennoi Voiny* (Moscow: Izd-vo Sotsial' no-ekonomicheskoi Literatury, 1960).
Skvortsov, L. I., *Tseny i Tsenoobrazovanie v SSSR* (Moscow: Vysshaya Shkola, 1972).
Sovetskaya Ekonomika v Period Velikoi Otechestvennoi Voiny, pod. red. I. A. Gladkova (Moscow: Nauka, 1970).
Sovetskaya Sotsialisticheskaya Ekonomika 1917–1957 gg (Moscow: 1957).
Suprunenko, N. I., *Ukraina v Veliloi˙ Otechestvennoi Voine Sovetskogo Soyuza (1914–1945 gg.)* (Kiev, 1956).
Tel'pukhovskii, V. B., in *Istoriya SSSR*, 1960, no. 6, pp. 27–42.
Tel'pukhovskii, V. B., *Osnovnye Periody Velikoi Otechestvennoi Voiny (Moscow: Mysl', 1965)*.
Volodarskii, L. M., *Promyshlennaya Statistika* (Moscow: Gosstatizdat, 1954).
Zagitov, B. Z., 'Nekotorye Voprosy Partiinogo Stroitel'stva v Pervye Poslevoennye gody (1945–50)', in *Voprosy Partiinoi Raboty na Sovremenoi Etape* (Moscow: Mysl', 1969).
Zverev, A. G., *Zapiski Ministra* (Moscow: Politizdat, 1973).

4 Sources Published in the USSR 1945–53

(a) Books and major articles

Arakelyan, A., *Industrial Management in the USSR*, translated by E. L. Raymond (Washington DC: Public Affairs Press, 1950).

Arakelyan, A., *Osnovnye Zadachi Poslevoennoi Pyatiletki* (Moscow: Gospolitizdat, 1946).

Arakelyan, A., *Vedushaya Rol 'Promyshlennosti v Razvitii Narodnogo Khozyaistva SSSR* (Moscow: Pravda, 1951).

Belov, P. A., *Voprosy Ekonomiki v Sovremmenoi Voine* (Moscow: Voenizdat, 1951).

Borisov, A. P., and Serov, K. P., *Leningradskaya Oblast' v Novoi Pyatiletke* (Leningradskoe Gazetno-Zhurnal'noe i Knizhnoe Izd-vo, 1947).

Chernyak., *O Tempakh Razvitiva Sotsialisticheskoi Promyshlennosti* (Moscow: Gospolitizdat, 1948).

Degtyar', D. D., *Pyatiletnii Plan Vosstanovleniya i Razvitiya Narodnogo Khozyaistva RSFSR Na 1946–1950 gg* (Ufa: Bashgosizdat, 1964).

Degtyar', D., *Vozrozhdenie Raionov RSFSR podvergavshchikhsya Nemetskoi Okkupatsii* (Moscow: Gospolitizdat, 1947).

Fish, D., *Voprosy Truda i Zarabotnoi Platy v Sotsialisticheskoi Promyshlennosti* (Lenizdat, 1948).

Gatovskii, L., Ekonomicheskaya Pobeda Sovetskogo Soyuza v Velikoi Otechestvenni Voine (Moscow: Politizdat, 1946).

Kantor, L. M., 'Perebazirovaniye Premyshlennosti SSSR', *Zapiski Leningradskogo Planovogo Instituta*, Vypusk VI (Leningrad: 1947).

Khachaturov, T. S., *Osnovy Ekonomiki Zhelezdorozhnogo Transporta* tom 1 (Moscow: 1946).

Lerskii, I., *Vosproizvodstvo Osnovnykh Fondov Promyshlennosti SSSR v Uslovivakh Voiny* (Moscow: Gosplanizdat, 1945).

Lure, A., *Voprosy Ekonomiki Zhelezdorozhnogo Transporta* (Moscow: 1948).

Ocherki po Istorii Organov Sovetskoi GosudarstvennoiVlasti (Moscow: Gosyurizdat, 1949).

Organizatsiva Finantsirovaniya i Kreditovaniya Kapital'nykh Vlozhenii, pod. red. N. N. Rovinskogo (Moscow: Gosfinizdat, 1951).

Saratovskii Gosudarstvennyi Universitet im N. G. Chernyshevskogo: Nauchnava Kohferentsiya 1946 goda: Sektsiya Geograficheskikh Nauk (Saratov: 1947).

Sokolov, B., *Promyshlennoe Stroitel'stvo v gody Otechestvennoi Voiny* (Moscow: Gosplanizdat, 1946).

Sokolov, B., *CCCP v Lesakh Velikoi Stroiki* (Moscow: Moskovskii Rabochii, 1947).

Stalin, I. V., *Ekonomicheskie Problémy Sotsializma v SSSR* (Moscow: Politizdat, 1952).
Volodarskii, L., *Vozrozhdenie Raionov SSSR postradavshchikh ot Nemetskoi Okkupatsii* (Moscow: Gosplanizdat, 1946).
Vikentev, A., *Ocherki Razvitiya Sovetskoi Ekonomiki v Chetvertoi Pyatiletke* (Moscow: Politizdat, 1952).
Voznesenskii, N. A., *Five Year Plan for the Rehabilitation and Development of the National Economy of the USSR 1946–50* (Moscow: 1946).
Voznesenskii, N. A., *Voennaya Ekonomika SSSR v Period Otechestvennoi Voiny* (Moscow: Gospolitizdat, 1948).
Zotov, V. P., *Razvitie Pishchevoi Premyshlennosti v Novoi Pyatiletki* (Moscow: Gospolitizdat, 1947).

(b) Newspapers and periodicals

(i) Dailies

Izvestiya.
Pravda.
Trud.

(ii) Other periodicals

Bol'shevik (later *Kommunist*).
Izvestiya Akademii Nauka: Otdel Ekonomiki i Pravo.
Planovoe Khozyaistvo.
Stál' (Organ Minchermeta SSSR).
Trudy Leningradskogo Finansovo-Ekonomicheskogo Instituta, vypusk 3 (Leningrad).
Tsvetnye Metally (Organ Mintsvetmeta SSSR).
Uchenie Zapisksi Leningradskogo Gosudarstvennogo Pedagogicheskogo Instituta im. A. I. Gertsena, kafedra tkonomicheskoi Geografii i kafedra Metodiki Geografii (Leningrad).
Vneshnaya Torgovlya.
Voprosy Ekonomiki (from 1947).
Voprosy Geografii (Sbornik 6–oi: 10–oi).

5 Primary Sources

(a) Statistical handbooks and other sources

Bolshaya Sovetskaya Entsiklopediya Tom 'SSSR' (Moscow).

Kapital'noe Stroitel'stvo v SSSR (Moscow: 1961).

'KPSS v Tsifrakh' (*Partiinaya Zhizn* 1967 no. 19 and 1973 no. 14)

Kuz'menko D. G., *Razvitie Sovetskoi Ekonomiki (Statisticheskie Materialy)* (Moscow, 1946).

Narodnoe Khozyaistvo SSSR (annually 1956–) (Moscow: 1957–).

Pinkhenson D. M., *Ekonomicheskaya Geografiya v Tsifrakh* (Moscow: Prosveshchenie, 1970).

Promyshlennost' SSSR: Statisticheskii Sbornik (Moscow, 1957 & 1964).

Strany Mira (Moscow, 1946).

Trud v SSSR (Moscow, 1967).

(b) Documents

Bor'ba KPSS za Vosstanovlenie i Razvitie Narodnogo Khozyaistva v Poslevoennoi Period: 1945–1953 gody: Dokumenty i Materialy (Moscow: Gospolitizdat, 1961).

Gruliow L. (ed.), *Current Soviet Policies: Documentary Record of 19th Party Congress*, (New York: Praeger, 1953).

Kommunisticheskaya Partiya v Period Velikoi Otechestvennoi Voiny: Dokumenty i Materialy (Moscow: Politizdat, 1961).

KPSS v Resolutsiyakh i Resheniyakah, tom 6 (Moscow: Politizdat, 1971).

Resheniya Partii i Pravïtel'stva po Khozyaistvennym Voprosam: tom 2, 1929–40 gody (Moscow: Politizdat, 1967); and tom 3, 1941–52 gody (Moscow: Politizdat, 1968).

Sbornik Soobshchenii Chrezvychainoï Gosudarstvennoi Kommissii o Zlodeyaniyakh Nemetsko – Fashistkikh Zakhvatchikov (Moscow: Politizdat 1946).

Index